China's Military Power

China's Military Power

Department of Defence, USA

Manas Publications
New Delhi-110002 (INDIA)

MANAS PUBLICATIONS
(We Convert Fighters into Writers)
(Publishers, Distributors, Importers & Exporters)
4402/5-A, Ansari Road (Opp. HDFC Bank)
Darya Ganj, New Delhi-110 002 (INDIA)
Off. ☎ 23260783, 23265523, Res. 23842660
Fax: 011-23272766
E-mail: manaspublications@gmail.com
manaspublications@yahoo.com

ISBN 978-81-7049-364-8
₹ 595/-

Designed at
Manas Publications

Printed in India at
HS Offsets, Delhi
and published by Mrs Suman Lata for
Manas Publications,
4402/5-A, Ansari Road,
(Opp. HDFC Bank), Darya Ganj,
New Delhi-110 002 (INDIA)

Contents

	Executive Summary	7
	Glossary of Acronyms	11
	China's Evolving Military Capability	15
1.	Understanding China's Strategy	21
2.	China's Military Strategy and Doctrine	39
3.	Force Modernization: Goals and Trends	59
4.	Resources for Force Modernization	77
5.	Force Modernization and Security in the Taiwan Strait	95
6.	Annual Update	105
7.	Special Topic: China's Global Military Engagement	123
	Appendix: China and Taiwan Forces Data	135
	Index	141

Executive Summary

China's rapid rise as a regional political and economic power with growing global influence has significant implications for the Asia-Pacific region and the world. The United States welcomes the rise of a stable, peaceful, and prosperous China, and encourages China to participate responsibly in world affairs by taking on a greater share of the burden for the stability, resilience, and growth of the international system. The United States has done much over the last 30 years to encourage and facilitate China's national development and its integration into the international system. However, much uncertainty surrounds China's future course, particularly regarding how its expanding military power might be used.

The People's Liberation Army (PLA) is pursuing comprehensive transformation from a mass army designed for protracted wars of attrition on its territory to one capable of fighting and winning short-duration, high-intensity conflicts along its periphery against high-tech adversaries - an approach that China refers to as preparing for "local wars under conditions of informatization." The pace and scope of China's military transformation have increased in recent years, fueled by acquisition of advanced foreign weapons, continued high rates of investment in its domestic defence and science and technology industries, and far-reaching organizational and doctrinal reforms of the armed forces. China's ability to sustain military power at a distance remains limited, but its armed forces continue to develop and field disruptive military technologies, including those for anti-access/area-denial, as well as for nuclear, space, and cyber warfare, that are changing regional military balances and that have implications beyond the Asia-Pacific region.

The PLA's modernization vis-à-vis Taiwan has continued over the past year, including its build-up of short-range missiles opposite the island. In the near-term, China's armed forces are rapidly developing coercive capabilities for the purpose of deterring Taiwan's pursuit of *de jure* independence. These same capabilities could in the future be used to pressure Taiwan toward a settlement of the cross-Strait dispute on Beijing's terms while simultaneously attempting to deter, delay, or deny any possible U.S. support for the island in case of conflict. This modernization and the threat to Taiwan continue despite significant reduction in cross-Strait tension over the last year since Taiwan elected a new president.

The PLA is also developing longer range capabilities that have implications beyond Taiwan. Some of these capabilities have allowed it to contribute cooperatively to the international community's responsibilities in areas such as peacekeeping, humanitarian assistance and disaster relief, and counter-piracy. However, some of these capabilities, as well as other, more disruptive ones, could allow China to project power to ensure access to resources or enforce claims to disputed territories.

Beijing publicly asserts that China's military modernization is "purely defensive in nature," and aimed solely at protecting China's security and interests. Over the past several years, China has begun a new phase of military development by beginning to articulate roles and missions for the PLA that go beyond China's immediate territorial interests, but has left unclear to the international community the purposes and objectives of the PLA's evolving doctrine and capabilities. Moreover, China continues to promulgate incomplete defence expenditure figures and engage in actions that appear inconsistent with its declaratory policies. The limited transparency in China's military and security affairs poses risks to stability by creating uncertainty and increasing the potential for misunderstanding and miscalculation. The United States continues to work with our allies and friends in the region to monitor these developments and adjust our policies accordingly.

Glossary of Acronyms

AAV	Amphibious Assault Vehicle
AEW&C	Airborne Early Warning and Control
AIP	Air Independent Propulsion
APEC	Asia-Pacific Economic Cooperation
ARATS	Association for Relations Across the Taiwan Strait
ARM	Anti-Radiation Missile
ASAT	Anti-Satellite
ASEAN	Association of Southeast Asian Nations
ASEM	Asia-Europe Meeting
ASBM	Anti-Ship Ballistic Missile
ASCM	Anti-Ship Cruise Missile
ASM	Air-to-Surface Missile
C4ISR	Command, Control, Communications, Computers, Intelligence, Surveillance and Reconnaissance
CAD/CAM	Computer-Assisted Design/Manufacturing
CCP	Chinese Communist Party
CMC	Central Military Commission
CNA	Computer Network Attack
CND	Computer Network Defence
CNE	Computer Network Exploitation
CNO	Computer Network Operations
CNP	Comprehensive National Power
CONUS	Continental United States

DDG	Guided-Missile Destroyer
DOD	Department of Defence
DTL	Defence Telephone Link
EEZ	Exclusive Economic Zone
EMP	Electro-Magnetic Pulse
EU	European Union
EW	Electronic Warfare
FBI	Federal Bureau of Investigation
FFG	Guided-Missile Frigate
FTA	Free Trade Agreement
GAD	General Armament Department
GDP	Gross Domestic Product
GEO	Geosynchronous Orbit
GPS	Global Positioning System
GSD	General Staff Department
HA/DR	Humanitarian Assistance/Disaster Relief
ICBM	Intercontinental-Range Ballistic Missile
ICE	U.S. Immigration and Customs Enforcement
IMF	International Monetary Fund
IOC	Initial Operational Capability
IRBM	Intermediate-Range Ballistic Missile
LACM	Land Attack Cruise Missile
LEO	Low Earth Orbit
MaRV	Manoeuvering Re-entry Vehicle
MINUSTAH	United Nations Stabilization Mission in Haiti
MIRV	Multiple Independently Targeted Re-entry Vehicles
MND	Ministry of National Defence
MR	Military Region
MRBM	Medium-Range Ballistic Missile
MRL	Multiple Rocket Launcher
NCO	Non-Commissioned Officer
NGO	Non-Governmental Organization

NM	Nautical Mile
OECD	Organization of Economic Cooperation and Development
OMTE	Outline of Military Training and Evaluation
OTH	Over-the-Horizon
PAP	People's Armed Police
PBSC	Politburo Standing Committee
PLA	People's Liberation Army
PLAAF	People's Liberation Army Air Force
PRC	People's Republic of China
SAM	Surface-to-Air Missile
SCO	Shanghai Cooperation Organization
SEF	Straits Exchange Foundation
SESS	Space Event Support Ship
SLBM	Submarine-Launched Ballistic Missile
SS	Diesel-Electric Attack Submarine
SRBM	Short-Range Ballistic Missile
SSBN	Nuclear-Powered Ballistic Missile Submarine
SSN	Nuclear-Powered Attack Submarine
UAV	Unmanned Aerial Vehicle
UCAV	Unmanned Combat Aerial Vehicle
UHF	Ultra High Frequency
UN	United Nations
UNCAR	United Nations Conventional Arms Register
UNFIL	United Nations Interim Force in Lebanon
USD	United States Dollars
ZANU-PF	Zimbabwe African National Union - Patriotic Front

China's Evolving Military Capability

Since the U.S. Department of Defence (DoD) submitted its first report pursuant to Section 1202 of the National Defence Authorization Act for Fiscal Year 2000, China has made considerable progress toward building and fielding credible and capable military power. The Department's understanding of China's military power has improved over the reporting period, but much remains to be learned about China's national and military strategies, progress and trends in its military modernization, and the related implications for regional security and stability. China has improved modestly the transparency of its military and security affairs, but until it begins to view transparency less as a transaction to be negotiated and more as a responsibility that accompanies the accumulation of national power, the insights reflected in this report will remain incomplete, bridged only by assessment and informed judgment. Several of these insights are highlighted below:

- **Defence Budget Outpacing Economic Growth** China's sustained economic growth, from a Gross Domestic Product (GDP) of $1.95 trillion in 2000 to a projected $4.19 trillion in 2008 (in 2008 USD) has enabled China to focus greater resources on building, equipping, and training the PLA without overwhelming the economy. One measure of increasing resourcing for the PLA is the official budget, which has more than doubled from $27.9 billion in 2000 to $60.1 billion in 2008 (in 2008 USD). The budget, however, does not capture the totality of military expenditure. Despite persistent efforts by the United States and others to gain greater clarity from China, or to improve estimates in the absence of such clarity, the Department of Defence's understanding of the resources, funding streams, and accounting mechanisms used to guide investment in the PLA has not improved measurably.

- **Strengthened Deterrent and Enhanced Strategic Strike:** Since 2000, China has shifted from a largely vulnerable, strategic deterrent based on liquid-fueled, intercontinental-range ballistic missiles (ICBMs) fired from fixed locations to a more survivable and flexible strategic nuclear force. The introduction of two new classes of ICBMs, the DF-31 and DF-31A, both road-mobile, solid-propellant systems (the latter capable of targeting any location in the continental United States) reflects this shift. While there is no evidence that China's doctrine of "no first use" has changed, the fielding of these forces, along with a projected new class of nuclear-powered ballistic missile submarine (SSBN)/ submarine-launched ballistic missile (SLBM) in 2009-2010 enabling a credible sea-based deterrent, will give China's leaders greater flexibility and options for strategic strike than previously available. While U.S. strategic forces still far outnumber those of China, China would be able to inflict significant damage on most large American cities with these survivable systems.

- **Improving Anti-Access/Area-Denial Capabilities** Since 2000, China has expanded its arsenal of anti-access and area-denial weapons, presenting and projecting increasingly credible, layered offensive combat power across its borders and into the Western Pacific. China has or is acquiring the ability to: 1) hold large surface ships, including aircraft carriers, at risk (via quiet submarines, advanced anti-ship cruise missiles (ASCMs), wire-guided and wake-homing torpedoes, or anti-ship ballistic missiles); 2) deny use of shore-based airfields, secure bastions and regional logistics hubs (via conventional ballistic missiles with greater ranges and accuracy, and land attack cruise missiles); and, 3) hold aircraft at risk over or near Chinese territory or forces (via imported and domestic fourth generation aircraft, advanced long-range surface-to-air missile systems, air surveillance systems, and ship-borne air defences). Advances in China's space-based reconnaissance and positioning, navigation, and timing, as well as survivable terrestrial over-the-horizon targeting, are closing gaps in the creation of a precision-strike capability.

- **Regional Conventional Strike:** Since 2000, China has continued its build-up of conventional ballistic missiles, building a nascent capacity for conventional short-range ballistic missile (SRBM) strikes against Taiwan into what has become one of China's primary instruments of coercion, not only of Taiwan but of other regional neighbours. In 2000, China's SRBM force was limited to one "regimental-sized unit" in southeastern China. China has expanded the force opposite Taiwan to seven brigades with a total of 1,050-1,150 missiles, and is augmenting these forces with conventional medium-range ballistic missile (MRBM) systems, such as the anti-ship ballistic missile, and at least two land attack cruise missile (LACM) variants capable of ground or air launch. Advanced fighters and bombers, combined with enhanced training for night time and overwater flights, provide the PLA with additional capabilities for regional strike or maritime interdiction operations.

- **Competing for Dominance of the Electromagnetic Spectrum:** The 2000 edition of this report observed that China is "working to ameliorate weaknesses in C4I training and plac[e] increased emphasis on 'electromagnetic warfare' to degrade or destroy enemy operational systems." At that time, the PLA's electronic warfare (EW) systems were derived mostly from a combination of "1950s-1980s technologies." By the 2006 edition of this report, China's investments in advanced EW programmes had given the PLA Air Force "technological parity with or superiority over most potential adversaries." By improving space-based and terrestrial C4ISR and by moving communications infrastructure to fiber, China is hardening its own capabilities while making gains in developing weapon systems (e.g., counterspace, computer network operations, and anti-radiation systems) to deny these capabilities to others. The 2004 introduction of the PLA concept of "local wars under conditions of informatization" has guided development in this area, positioning the PLA to contest electromagnetic dominance in the early phases of future campaigns.

- **Persistent Limitations:** As China's capabilities for local and regional operations have increased in certain areas since 2000, a number of limitations appear to have persisted. The PLA has developed new doctrine for joint warfighting and implemented organizational changes, such as including service commanders on the Central Military Commission, to facilitate the transition to a more "joint" force. However, joint integration still lags. Similarly, PLA air and amphibious lift capacity has not improved appreciably since 2000 when the Department of Defence assessed the PLA as capable of sealift of one infantry division. Likewise, China's current ability to deliver about 5,000 parachutists in a single lift (less if equipment is carried at the same time) is similar to previous assessments. China's at-sea replenishment has improved with experience since 2000, but the PLA Navy today remains limited by a small number of support vessels - much as it did then. In 2000, the Department of Defence projected aerial refueling as an operational capability by 2005. Today, while China has a few aerial refueling aircraft, it does not have the number of tankers, properly equipped combat aircraft, or sufficient training to employ this capability for power projection.

- **Shifting Dynamics in the Taiwan Strait:** Since 2000, there have been two peaceful political transitions on Taiwan and a gradual and steady maturation of Taiwan democracy. While Beijing's strategy toward Taiwan appears to have shifted from seeking an early resolution of the Taiwan issue to one of preventing Taiwan's *de jure* independence, by force if necessary, Beijing's objective of unifying Taiwan with the Mainland has not changed. Since 2000, the military balance in the Taiwan Strait has continued to shift in Beijing's favour, marked by the sustained deployment of advanced military equipment to the Military Regions opposite Taiwan. In the 2002 report, the Department of Defence assessed that Taiwan "has enjoyed dominance of the airspace over the Taiwan Strait for many years." This conclusion no longer holds true. With this reversal, China has been able to develop a range of limited military options to attempt to coerce Taipei.

CHAPTER 1

Understanding China's Strategy

"China has become an important member of the international system, and the future and destiny of China have become increasingly closely connected with the international community. China cannot develop in isolation from the rest of the world, nor can the world enjoy prosperity and stability without China."

"World peace and development are faced with multiple difficulties and challenges. Struggles for strategic resources, strategic locations and strategic dominance have intensified."

- China's National Defence in 2008

Overview

People's Republic of China (PRC) leaders have outlined national strategic objectives in a series of "white papers" on China's foreign policy, national defence, and arms control. However, such documents offer only limited insights on the motivations behind the objectives, or the specific strategies to achieve them. The linkages between the occasional strategic pronouncement and actual policy decisions in China are not apparent, especially during periods of crisis. As a result, the study of PLA views on grand strategy remains an inexact science. Still, it is possible to make some generalizations about China's strategy based on tradition, historical pattern, official statements and papers, and emphasis on certain military capabilities and diplomatic initiatives.

Leadership Guidance on Foreign and Security Policy

When analyzing China's foreign and security policy, Chinese strategists and analysts occasionally cite guidance from former paramount leader Deng Xiaoping in the early 1990s: "observe calmly; secure our position; cope with affairs calmly; hide our capacities and bide our time; be good at maintaining a low profile; and never claim leadership." This guidance reflected Deng's belief that China's foreign policy and security strategy had to reinforce its core national interest of promoting domestic development by avoiding foreign risk, high-profile international engagement or provocations, or pretenses of international leadership. Some analysts see Deng's guidance as prescribing deliberate efforts to conceal intentions and capabilities.

As China's global interests and influence have expanded in recent years, its diplomatic and military presence and engagement have become more visible and active to the world. PRC President Hu Jintao's own ideological formulation – "Harmonious World," which emphasizes "diversity" and "equality" in international relations along with traditional PRC foreign policy dictums of "noninterference" and the "democratization of international relations" – was endorsed at the 17th Party Congress in October 2007. Hu's "Harmonious World" formulation reflects an evolution in the general tone and conduct of China's foreign and security affairs in recent years, but likely has not completely superseded Deng's thinking.

Strategy with Chinese Characteristics

PRC strategy is one of maintaining balance among competing priorities for national economic development and sustaining the type of security environment within which such development can occur. China's leaders describe the initial decades of the 21st Century as a "20-year period of opportunity," meaning that regional and international conditions will generally be peaceful and conducive to China's rise to regional pre-eminence and global influence.

PLA leaders and strategists often discuss China's strategy in terms of building "comprehensive national power". Comprehensive national power (CNP) is the concept by which China's strategic planners use qualitative and quantitative variables to evaluate and measure China's standing in relation to other nations. CNP incorporates both soft, internally oriented indicators of strength (e.g., economic prosperity, domestic cohesion, and cultural influence) and hard, externally oriented measures (e.g., the size of a state's nuclear arsenal, territory, military capability, diplomatic influence, economic influence, and international prestige). As evidenced by the composition of CNP, there is a tendency among China's strategists to perceive a link between the internal and external dimensions of strength and weakness. This indicates that China's decision-makers might see internal turmoil as an invitation to hostile external forces or possibly as the work of such forces. China's leaders may also perceive external challenges as connected to domestic enemies.

Insights on China's Strategy and Priorities

China's leaders appear to have adopted a set of enduring strategic priorities, which include perpetuating

Chinese Communist Party (CCP) rule, sustaining economic growth and development, maintaining domestic political stability, defending China's national sovereignty and territorial integrity, and securing China's status as a great power. Less clear are the specific strategies and plans Beijing has developed to achieve these objectives, the decision-making structures that guide strategy development and execution, and the manner and direction in which these priorities may adjust in response to changes in the security environment.

Regime survival and the perpetuation of CCP rule shape the strategic outlook for China's leaders and drive many of their choices. As a substitute for the failure of communist ideology to unify the population and mobilize political support, Party leaders have relied on economic performance and nationalism as the basis for regime legitimacy. However, each contains risks that may undermine political control. For example, while China's leaders have stoked nationalist sentiment to manipulate public opinion, deflect domestic criticism, or bolster diplomacy, they are aware that protests can be difficult to control once begun and could easily turn against the state itself. Similarly, China's rapid economic growth - vital to the success of the CCP strategy - has led to increased economic inequality and dislocation, official corruption, and environmental degradation. The ongoing impact of the international financial downturn may provide an opportunity to gauge more clearly China's strategy and priorities, and to see if and how each is modified by this event.

China's leaders have reaffirmed and continue to support "reform and opening," which began in 1978 as the fundamental basis for China's overall strategy and policy. However, two central perceptions increasingly appear in senior PRC leadership statements and commentary, suggesting a growing recognition that the process of "reform and opening" has engendered several contradictions and challenges:

- First, reforms have enabled China to experience rapid growth in economic, political, and military power, but have also led to significant new challenges to internal stability. Official commentary on the 6th Plenum of the 16th CCP Party Congress in October 2006 stated that "economic restructuring and social transformation" produced a corresponding increase in "factors of uncertainty" in China's domestic stability.
- Second, reforms have increasingly propelled China into a global security environment in which external events can no longer be isolated from their effects on China's internal situation, and vice versa. Official PRC media and policy documents continue to assess that "peace, development, and cooperation have become the trend of the contemporary era… making the external environment generally favorable to our country's development." However, there is growing acknowledgement that external challenges, including regional and great power relations, cross-Strait relations, and non-traditional security threats have the potential to influence the Mainland's future.

These dual perceptions have led Party leaders to conclude that, through 2020, they should focus on managing or exploiting external tensions, especially with the great powers, to maintain an environment conducive to China's development. As former State Councilor and Minister of Foreign Affairs Tang Jiaxuan explained in November 2005:

"It is necessary to exert the greatest efforts to avoid becoming the focal point of major international conflict, reduce external pressures and obstructions to the development of our country, [and] create favorable external conditions for China to achieve its planned development goals."

This does not mean that China's leaders are exclusively inward-focused. Beijing's growing regional and, to a lesser extent, global economic stature partly drive a more active external posture in which it demonstrates a willingness to assert its interests while taking on a more active role in resolving disputes and promoting regional cooperation. Beijing continues to threaten the use of military force to compel settlement of the Taiwan dispute if force is deemed necessary, and to take steps to assert rights in the South China Sea that other regional countries do not accept. China has increased the scope of its participation in United Nations (UN) peace operations and continues to expand foreign military sales and training of foreign officers. These trends indicate that Beijing's methods of asserting and protecting its interests are increasingly diverse and that it is prepared to employ a variety of cooperative and coercive approaches to achieve its objectives.

Resource Needs as a Factor in China's Strategy

As China's economy grows, dependence on secure access to markets and natural resources, particularly metals and fossil fuels, has become an increasingly significant factor shaping China's strategic behaviour.

Although China is expected to continue to rely on coal as its primary fuel source, consumption of petroleum and other liquid fuels will likely grow significantly, due in large part to growth in the transportation sector. China plans to increase natural gas utilization from three percent to eight percent of total consumption by 2010. Similarly, China plans to build some 40 1,000-megawatt nuclear power reactors by 2020, increasing nuclear power from two to six percent of total electricity output.

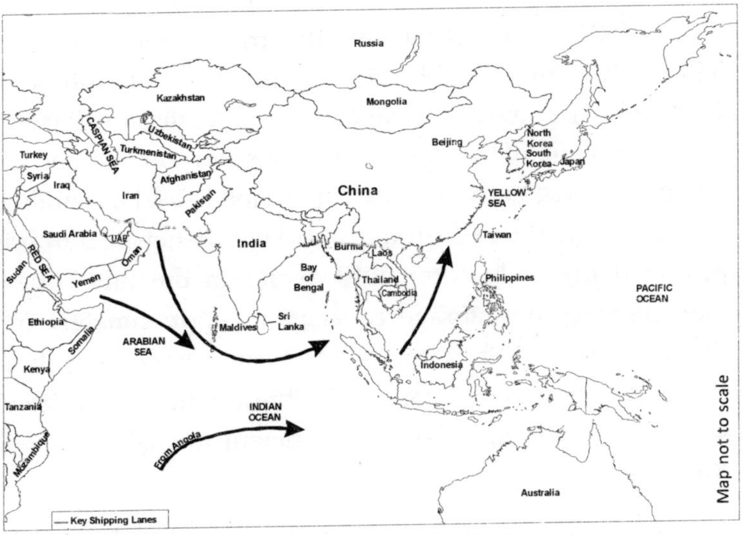

***China's Critical Sea Lanes.** Like many other industrialized East Asian countries, China is heavily dependent upon critical sea lanes for its energy imports. Over 80 percent of China's crude oil imports transit the Strait of Malacca.*

China currently consumes approximately 7.85 million barrels of oil per day. By 2015, China's oil consumption could rise to 10-12 million barrels per day. In May 2008, China became the world's second largest importer of crude oil, surpassing Japan. China currently imports over 53 percent of its oil (around 4.04 million barrels per day in 2007). China imports about 46 percent of its imports from the Middle East, 32 percent from Africa, and 5 percent from East Asia. In 2008, Saudi Arabia was China's largest supplier at about 725,000 barrels of oil per day. Saudi Arabia is followed by Angola at about 596,000 barrels of oil per day, and Iran at about 425,000 barrels of oil per day. The vast majority of its imported oil is carried on ships transiting through the Malacca or Lombok/Makkasar Straits.

In 2004, China began building a strategic petroleum reserve. It completed the construction of the first phase in 2008, which provides a capacity of 100 million barrels, the equivalent of 25 days of China's current annual net oil imports. The second phase is planned to add 200 million barrels, for a total of around 42 days of net oil imports projected for 2015. After 2010, work on the third phase may increase net storage capacity to approximately 500 million barrels. But, without significant improvements to China's transportation and distribution networks, gross storage capacity may prove insufficient to cushion severe disruptions.

In the last decade, China has pursued long-term supply contracts with a diverse range of supplier nations including Chad, Egypt, Indonesia, Kazakhstan, Nigeria, Oman, Russia, Australia, Saudi Arabia, Sudan, and

Venezuela. In 2006, China's top three suppliers were: Angola (16 percent), Saudi Arabia (16 percent), and Iran (12 percent). In 2007, six percent of China's crude oil imports came from Sudan. Currently, slightly over half of China's imported oil comes from the Middle East and almost a quarter is imported from Africa.

China's Territorial Disputes

China's use of force in territorial disputes has varied widely throughout history. Some disputes led to war, such as China's border conflicts with India in 1962 and Vietnam in 1979. A contested border with the former Soviet Union during the 1960s raised the possibility of nuclear war. In more recent cases, China has been willing to compromise with and even offer concessions to its neighbours. Since 1998, China has settled eleven land territorial disputes with six of its neighbours. Several disputes continue over exclusive economic zones (EEZs) and ownership of potentially rich, off-shore oil and gas deposits.

The East China Sea contains approximately 7 trillion cubic feet of natural gas and up to 100 billion barrels of oil. Japan maintains that an equidistant line from each country involved should separate the EEZs, while China claims an Extended Continental Shelf beyond the equidistant line to the Okinawa Trench (which almost reaches Japan's shore). In June 2008, China and Japan signed an agreement to temporarily shelve the EEZ dispute and develop jointly the Chunxiao/Shirakaba gas field. China and Japan continue to dispute possession of the nearby Senkaku Islands. In December 2008, two PRC surface ships intruded on waters surrounding the islands, but withdrew after Tokyo lodged an official protest with Beijing.

On December 31, China and Vietnam signed a treaty demarcating their land borders. The treaty followed a ten-year process of implementing a 1999 agreement. However, the treaty did not address the sea border, which remains disputed.

The South China Sea plays an important role in Northeast Asia and Southeast Asia security considerations. Northeast Asia relies heavily on the flow of oil through South China Sea shipping lanes, which provide 80 percent of the crude oil to Japan, South Korea, and Taiwan. China claims sovereignty over the Spratly and Paracel island groups -- claims disputed in whole or part by Brunei, the Philippines, Malaysia, Indonesia, and Vietnam. Taiwan, which occupies Itu Aba in the Spratly Islands, also claims all four island groups in the South China Sea. In December 2007, China announced the establishment of a city administration, "Sansha City," to assert "indisputable sovereignty" and jurisdiction over the islands of the South China Sea "and the adjacent waterways," prompting street protests in Vietnam. In October 2008, China and Vietnam released a joint statement that pledged to "gradually advance the negotiations on demarcation of these maritime zones and ... jointly exploit the zones."

While China and India have improved bilateral relations, tensions remain along their shared 4,057 km border, most notably over Arunachal Pradesh. In June 2008, PRC troops entered more than a kilometer into the northernmost point of India's Sikkim state. PRC soldiers, in May, had threatened to demolish stone structures in the area. In May 2008, India's Minister of External Affairs Pranab Mukherjee visited the PRC and signed a Memorandum of Understanding to expand bilateral military cooperation. Following local elections in November 2008, Mukherjee stated that Arunachal Pradesh is "an integral part of India." A spokesman for the PRC Ministry of Foreign Affairs rejected the statement.

China has also pursued equity positions in a variety of overseas energy assets and investments, although these remain small compared to PRC demand and investments by the international oil majors. China's national oil companies have invested in oil ventures (oilfield development, and pipeline and refinery projects) in Kazakhstan, Turkmenistan, Nigeria, Sudan, and in over 20 other countries in North Africa, Central Asia, Southeast Asia, Latin America, and North America.

Factors Shaping Pathways to China's Future

Since initiating "reform and opening" in 1978, China has made tremendous economic progress and has overcome many developmental challenges. In 30 years, these reforms have lifted hundreds of millions of its citizens out of poverty, improved domestic stability, expanded China's interconnections with the rest of the world, and increased China's influence in international affairs. China continues to face many problems, but the accomplishments of China's leaders in these areas are impressive. The United States welcomes the rise of a peaceful and prosperous China.

There are forces - some beyond the control of China's leaders - that could reinforce a relatively inward focus on national strategy, but could also divert China from a peaceful pathway. Which pathway China pursues, or finds itself upon, will be determined in large part by the choices China's leaders make. These choices are influenced by a set of drivers and inhibitors that will both enable and constrain their ability to achieve their objectives.

Economics: Continued economic development, central to China's emergence as a regional and global power, remains the foundation of the Party's popular legitimacy and underwrites its military power. Since 1978, China's economic growth has improved the quality of life of its citizens, has garnered support for the Party, and has contributed to regional and global economic growth. In contrast, economic shocks, like the current global financial crisis, might place political stress on the Chinese system. As many China analysts have noted, underlying weaknesses (e.g., undervalued currency, non-performing loans, inefficient state-owned enterprises, and economic disparity between urban and rural areas) threaten continued economic growth. Economic shocks, setbacks, or even modestly slower growth could lead to higher unemployment, inflation, and significant unrest, potentially giving rise to greater reliance on nationalism to maintain popular support for the Party. Unexpected increases in resource demand, global resource shortages or price shocks, restricted access to resources, or shrinking demand for labour and manufacturing, could also impact China's strategic outlook and behaviour, and might force China's leadership to re-examine its resource allocation priorities, including those for the military.

Demographic Pressures: Demographic stresses will increase in the future, creating a structural constraint on China's ability to sustain high growth rates. Between 2000 and 2030, over 400 million people - a population greater than the entire United States - will transition from the countryside into urban areas. As a result of this shift, China's leaders will face challenges in terms of job creation as well as satisfying housing and infrastructure needs. Some estimates suggest that China is expected to account for half of global building construction during that period. China's population is also aging rapidly; its population of 146 million senior citizens will increase to an estimated

290 million by 2025. Accommodating the needs of a large senior citizen population will present challenges to the CCP's ability to maintain economic growth, growing defence budgets, and perhaps domestic stability.

Domestic Political Pressures: A successful Beijing Olympics and perceptions of a competent PLA response to the Sichuan earthquake gave Party leaders and the PLA a boost in the eyes of the public. China's state-run media, for example, popularized images of Premier Wen Jiabao's trip to Sichuan to survey the damage and console the victims. However, reports of shoddy construction and corruption related to schools that collapsed in Sichuan as a result of the earthquake soon elicited an outpouring of popular criticism. The Party's decision to suppress reports of melamine in China's milk during the Olympics produced additional public condemnation. More broadly, Party leaders are increasingly confronted with popular demands for improved government responsiveness and accountability. The public debate, including in online forums, regarding the melamine issue and the government's delay in publicizing it are some evidence of greater public space for discussion of such issues. However, the government continues to limit sharply the scope of such debates. How China's leaders choose to address these challenges will affect their ability to maintain domestic stability and their monopoly on political power.

While preserving one-party rule, Beijing has enacted administrative reforms and expanded avenues for expert - and occasionally public - input as evidenced by the emergence of non-governmental organizations (NGOs) throughout China focused on addressing the concerns of the population. China's leaders have also launched a major personnel reform programme aimed at improving communication between central leaders, local cadres, and urban and rural workers with grievances. A stated primary objective of this programme is to defuse "inner contradictions among the people."

However, overt acts of dissent remain criminalized, media and the internet are tightly controlled, independent trade and labour unions are suppressed, ethnic Tibetan and Uighur minorities are repressed, and religious groups not recognized by the regime continue to be harassed. The Party is wary of unsanctioned organizations in China, even if non-political, fearing these organizations could facilitate organized opposition.

Corruption: Despite efforts to curb official corruption in China, it remains pervasive, structural, and persistent due, in part, to the high degree of state involvement in the economy and the weakness of the rule of law.

In June 2008, the CCP Central Committee announced a 5-year anti-corruption campaign. Several procuratorates used mass text messaging with reward offers starting at $440 for tips leading to a conviction, which led to a doubling of corruption investigations within weeks. Also in 2008, dozens of officials were removed from their posts for misconduct associated with earthquake relief efforts.

In 2001, 65 percent of embezzlement cases involved multiple officials, indicating the activity of independent networks of elites colluding at the expense of the state. China's National Audit Agency uncovered $170 billion of misappropriated and misspent public funds between 1996 and 2005, and academic research estimates that the direct costs of corruption in 2003 amounted to as much as $86 billion (three percent of GDP), an amount that was more than double China's announced defence budget for that year. Corruption directly affects the PLA; bribery for advancement and promotion, unauthorized contracts and projects, and weapons procurement are all identified by the PLA as corruption problems.

Beijing's response has focused on the use of criminal prosecution to deter illicit behaviour. Half of provincial transportation chiefs in China have been sentenced to jail terms (some have been executed) for corruption. In July 2007, the Director of China's Food and Drug Administration was executed for taking bribes to approve fake drugs.

Environment: China's economic development has come at a significant environmental cost. Acceptable air quality for the Beijing Olympics could only be achieved by shutting down much of nearby industry and local traffic. A 2007 World Bank report estimated that air and water pollution costs China almost six percent of its GDP annually, and contributes to increasing levels of cancer, lung disease, and other ailments. China's leaders are concerned that these environmental problems could undermine regime legitimacy by threatening China's economic development, public health, social stability, and international image. In 2006, China's top environmental official, Zhou Shengxian, announced that there had been 51,000 pollution-related protests in 2005 (almost 1,000 per week). Pollution and deforestation in China have worldwide implications. China may have overtaken the United States as the world's largest emitter of atmospheric carbon dioxide. Japan and South Korea both suffer from acid rain produced by China's coal-fired power plants and yellow dust storms that originate in the Gobi desert.

Cross-Strait Dynamics: Despite a reduction in tensions following the March 2008 election and May 2008 inauguration of Taiwan President Ma Ying-jeou, a potential military confrontation with Taiwan and the prospect of U.S. military intervention remain the PLA's most immediate military concerns. China's current strategy toward Taiwan appears to be one of preventing any moves by Taipei toward de jure independence, rather than seeking near-term resolution. A perceived shift in military capabilities or political will on either side, or a change in the internal political landscape on Mainland China or Taiwan, could cause Beijing to calculate its interests, and its preferred course of action differently.

Regional Concerns: With China close to, or an interested party in, many of the world's "flashpoints" (e.g., Taiwan, North Korea, the Spratly Islands, the Senkaku Islands, Afghanistan, and Pakistan), China's leaders hope to avoid regional instability spilling across China's borders and thereby interfering with economic development or domestic stability. Changes in regional security dynamics could lead to shifts in China's military development and deployment patterns, likely with consequences for neighboring states. Examples of such changes from Beijing's perspective include disruptions on the Korean Peninsula (e.g., a North Korean collapse); democratic revolutions in Central Asia, which Beijing would perceive as posing near-term and long-term security challenges; a downturn in relations with Japan; and perceived threats to China's ability to access foreign resources and transport them back to China.

Looking to the Future

China's current strategy is to manage external tensions to assure an environment that is conducive to economic development. This strategy appears to be accepted widely by Beijing's foreign and security policy establishment. However, differences of opinion within China occasionally surface, particularly in academic circles, about how China can achieve these goals and how it can best prevent conflict with its neighbours and the United States over time. Some prefer the traditional strategy of Deng Xiaoping: avoid leadership roles and seek to avoid having China being burdened with excessive international responsibilities by those who are trying to encourage China to play a more active and constructive role in addressing regional and international problems. Another group believes that such passivity is untenable as China's power grows. This group asserts that China should work actively to cooperate with regional actors and the United States not only to increase Chinese influence, but also to reassure neighbours and more distant great powers that China's rise will not pose a destabilizing threat to their security. Still others believe that China needs to be tougher and more assertive in protecting its interests by countering perceived efforts by the United States to bully China or constrain its influence in relation to actors such as Taiwan, Japan, South Korea and Southeast Asia.

These discussions, however, appear to remain largely on the margins. Absent a perceived challenge to the legitimacy and authority of the CCP, a fundamental shift in China's strategy is unlikely to occur prior to the 2012 18th Party Congress, which should install the next, and fifth, generation of Party leaders.

CHAPTER 2

China's Military Strategy and Doctrine

"The army must provide an important force guarantee for the Party to consolidate its governance status, provide a powerful security guarantee for safeguarding the period of important strategic opportunity for national development, provide effective strategic backing for the safeguarding of national interests, and play an important role in preserving world peace and promoting common development."

- PRC President Hu Jintao

Overview

PLA theorists have developed a framework for doctrine-driven reform to build a force capable of fighting and winning "local wars under conditions of informatization." Drawing upon foreign military experiences, particularly US-led campaigns up to and including Operation Enduring Freedom and Operation Iraqi Freedom, Soviet and Russian military theory, and the PLA's own combat history, China is transforming across the whole of its armed forces.

The pace and scale of these reforms are broad and sweeping. However, the PLA remains untested in modern combat. This lack of operational experience complicates outside assessment of the progress of China's military transformation. The same applies to China's internal assessments of its own military capabilities, in which China's civilian leaders must rely upon the advice of operationally inexperienced commanders or from "scientific" combat models divorced from the realities of the modern battlefield.

Military Strategic Guidelines

China does not publish equivalents to the US National Security Strategy, National Defence Strategy, or National Military Strategy. Outside observers therefore have few direct insights into the strategic concepts motivating China's force build-up, the leadership's thinking about the use of force, and contingency planning that shapes the PLA's force structure or doctrine. Analysis of authoritative speeches and documents suggests China relies on a body of overall principles and guidance known as the "National Military Strategic Guidelines for the New Period" to plan and manage the development and use of the armed forces. While referenced and summarized in the 2008 Defence White Paper, the PLA has not made the contents of the "Guidelines" available for outside scrutiny.

Academic research suggests that the current "Guidelines" most likely date to 1993, reflecting the impact of the 1991 Persian Gulf War and the collapse of the Soviet Union on PRC military-strategic thinking. Recent revisions to the "Guidelines" likely reflect updates of China's perceptions of its security environment and the character of modern war, integrate lessons learned from China's military modernization, and emphasize building forces to win "local wars under conditions of informatization."

The operational, or "active defence," component of the "Guidelines" posits a defensive military strategy in which China does not initiate wars or fight wars of aggression, but engages in war only to defend national sovereignty and territorial integrity. Once hostilities have begun, according to the PLA's *Science of Campaigns* (2000), "the essence of [active defence] is to take the initiative and to annihilate the enemy.... While strategically the guideline is active defence, [in military campaigns] the emphasis is placed on taking the initiative in active offence. Only in this way can the strategic objective of active defence be realized.

Informatization

The concept of "informatization" emphasizes the effects of modern information technology on military decision and weapons employment cycles. The term officially entered the PLA's lexicon in 2002 when then-CCP General Secretary and Central Military Commission (CMC) Chairman Jiang Zemin referred to the concept as necessary for the PLA's rapid modernization and for enabling Integrated Joint Operations in a speech before the 16th Party Congress. Jiang's address recognized that moving China's military on a path toward informatization would require integrating the entire PLA with common information systems as well as forcing a new organizational model for warfighting. The PLA formally institutionalized the concept in 2004. Recent PLA analyses of U.S. and coalition operations in Iraq and Afghanistan have re-emphasized the importance of informatization and joint operations. For example, PRC official media cited an analyst from China's National Defence University as saying that "the greatest lesson learned from the war [in Iraq] was the importance of information warfare."

The PLA is attempting the concurrent pursuit of 'mechanization' (equipment acquisitions) and 'informatization' (networking of equipment). As a consequence, and in recognition of the high costs of force-wide refitting with state-of-the-art weapons systems, the PLA is selectively acquiring new generation technologies in some areas while deferring new acquisitions in others in favour of upgrading older, but capable, systems for networked operations.

PLA studies of informatization and observations of recent US and Coalition military operations in Iraq and Afghanistan have informed an ongoing debate within the PLA about the PLA's traditional ground combat focus on attrition and annihilating the enemy. A new model using "information-plus-firepower" considers ground forces as integrated within a joint force focused on rapid occupation of key strategic targets and on stabilization of the battlefield. It is unlikely that this debate has influenced authoritative PLA operational guidelines, but it may do so in the future.

In addition to developing the capacity to "annihilate" opposing forces, the PLA is exploring options for limited uses of force. PLA campaign theory defines these options to include "non-war" uses of force - an extension of political coercion. The 1995 and 1996 amphibious exercises and missile firings in the Taiwan Strait are examples of "non-war" uses of force. Other options for limited uses of force include air and missile strikes, targeted attacks against adversary leaders and sabotage.

The PLA is developing and implementing a supporting doctrine for "active defence" warfare and new operational methods across the various services.

Naval Warfare: The naval component of "active defence" is termed "Offshore Active Defence." The PLA Navy has three main missions: resist seaborne aggression, protect national sovereignty, and safeguard maritime rights. PLA Navy doctrine for maritime operations focuses on six offensive and defensive campaigns: blockade, anti-sea lines of communication, maritime-land attack, anti-ship, maritime transportation protection, and naval base defence.

Ground Warfare: Under "active defence," ground forces are tasked to defend China's borders, ensure domestic stability, and exercise regional power projection. PLA ground forces are transitioning from a static defensive force allocated across seven internal MRs - oriented for positional, mobile, urban, and mountain offensive campaigns; coastal defence campaigns; and landing campaigns - to a more offensive and manoeuvre-oriented force organized and equipped for operations along China's periphery. China's ground forces are placing emphasis on integrated operations (especially with aviation forces), long-distance mobility, "quick tempo" operations, and special operations. Their reforms are modeled on Russian doctrine and US military tactics. The ground forces appear to be leading the PLA's effort to experiment with ad hoc, multi-service, joint tactical formations to execute integrated joint operations.

Offense as Defence

Beijing's definition of an attack against its sovereignty or territory is vague. The history of modern Chinese warfare provides numerous case studies in which China's leaders have claimed military preemption as a strategically defensive act. For example, China refers to its intervention in the Korean War (1950-1953) as the "War to Resist the United States and Aid Korea." Similarly, authoritative texts refer to border conflicts against India (1962), the Soviet Union (1969), and Vietnam (1979) as "Self- Defence Counter Attacks." This logic suggests the potential for China to engage in military preemption, prevention, or coercion if the use of force protects or advances core interests, including territorial claims (e.g., Taiwan and unresolved border or maritime claims).

Chinese strategic-level military theory establishes seemingly contradictory guidance: "strike only after the enemy has struck," and "seize the initiative." Of note, China's 2008 Defence White Paper features a slightly different construction:

"Strategically, [the PLA] adheres to the principle of...striking and getting the better of the enemy only after the enemy has started an attack".

Yet, the authoritative work, *Science of Military Strategy*, makes it clear that the definition of an enemy strike is not limited to conventional, kinetic military operations. Rather, an enemy "strike" may also be defined in political terms. Thus:

"Striking only after the enemy has struck does not mean waiting for the enemy's strike passively...It doesn't mean to give up the "advantageous chances" in campaign or tactical operations, for the 'first shot' on the plane of politics must be differentiated from the 'first shot' on that of tactics...

If any country or organization violates the other country's sovereignty and territorial integrity, the other side will have the right to 'fire the first shot' on the plane of tactics."

These passages illustrate the ambiguity of PRC strategic thinking as well as the justification for offensive - or preemptive - military action at the operational and tactical level under the guise of a defensive posture at the strategic level.

PLA Reserve Forces and China's Militia

The PLA reserve is a key component of China's national defence. During peacetime, the National Defence Reserve conducts training and maintains social stability. During wartime, PLA reserve units may be transferred to active duty as directed by national mobilization orders. In 2008, the total strength of the PLA reserve was estimated at over 500,000. The PLA reserve is striving to become a more professional force by strengthening recruitment, training, and infrastructure. Professionalization will also require more emphasis on the development of PLAAF, Navy, and Second Artillery Corps reserve units and combat support units, which will probably come at the expense of traditional ground combat units.

China's militia system provides a large pool of personnel for national defence. All PRC males between the ages of 18 and 35 not currently serving in the military are technically part of the militia system. Many members of the militia do not belong to organized units, and China's 2004 Defence White Paper claimed that ten million people were organized into militia units. China's 2008 Defence White Paper indicates that these numbers will decrease to 8 million by the end of the 11th Five Year Plan (2006-2010). During wartime, militia may be mobilized to support the war effort within their home province. The functions of militia vary from locality to locality, covering diverse tasks such as air defence, emergency response, and technical support which can include technical maintenance and repair, as well as computer network operations.

Air Warfare: The PLA Air Force (PLAAF) is converting from a limited territorial defence force to a more flexible and agile force able to operate off-shore in both offensive and defensive roles, using the US and Russian air forces as models. Mission focus areas include: strike, air and missile defence, early warning and reconnaissance, and strategic mobility. The PLAAF also has a leading role in the "Joint Anti-Air Raid" campaign, which appears to form the basis for much of China's planning for anti-access and area-denial operations. Underscoring the ambiguity of offense and defence in PLA theory, the Joint Anti-Air Raid campaign is strategically defensive in nature, but at the operational and tactical levels, it calls for attacks against adversary bases and naval forces.

Space Warfare: PLA strategists see space as central to enabling modern informatized warfare; indeed, a 2003 analytic article in the PLA's leading journal was titled "Control of Space is Decisive in Modern High-Tech Informatized Warfare." That said, China does not appear to have a dedicated space campaign; rather, space operations form an integral component of all campaigns. The PLA's military theoretical journal *China Military Science* argues that "it is in space that information age warfare will come into its more intensive points." Specifically, space-based Command, Control, Communications, Computers, Intelligence, Surveillance, and Reconnaissance (C4ISR) is important to enable and coordinate joint operations and win modern wars. Accordingly, the PLA is acquiring technologies to improve China's space-based C4ISR. A PLA analysis of U.S. and Coalition military operations reinforced the importance of operations in space to enable informatized warfare, claiming that "space is the commanding point for the information battlefield. Battlefield monitor and control, information communications, navigation and position, and precision guidance all rely on satellites and other sensors."

Concurrently, China is developing the ability to attack an adversary's space assets. PLA writings emphasize the necessity of "destroying, damaging, and interfering with the enemy's reconnaissance/observation and communications satellites," suggesting that such systems, as well as navigation and early warning satellites, could be among initial targets of attack to "blind and deafen the enemy." The same PLA analysis of U.S. and Coalition military operations also states that "destroying or capturing satellites and other sensors ... will deprive the opponents of initiatives on the battlefield and [make it difficult] for them to bring their precision guided weapons into full play."

PRC military writings also discuss the importance of space warfare for its supposed psychological impact on the will of the adversary to fight. In a PLA National Defence University book, *Joint Space War Campaigns (2005)*, author Colonel Yuan Zelu writes:

"[The] goal of a space shock and awe strike is [to] deter the enemy, not to provoke the enemy into combat. For this reason, the objectives selected for strike must be few and precise ... [for example], on important information sources, command and control centers, communications hubs, and other objectives. This will shake the structure of the opponent's operational system of organization and will create huge psychological impact on the opponent's policymakers."

The January 2007 test of a direct ascent anti-satellite (ASAT) weapon demonstrates that the PLA's interest in counterspace systems is more than theoretical. In addition to the "kinetic kill" capability demonstrated by the ASAT test, the PLA is developing the ability to jam, blind, or otherwise disable satellites and their terrestrial support infrastructure.

Integrated Network Electronic Warfare: PRC military writings highlight the seizure of electromagnetic dominance in the early phases of a campaign as among the foremost tasks to ensure battlefield success. PLA theorists have coined the term "integrated network electronic warfare" to describe the use of electronic warfare, computer network operations, and kinetic strikes to disrupt battlefield network information systems that support an adversary's warfighting and power projection capabilities. PLA writings on future models of joint operations identify "integrated network electronic warfare" as one of the basic forms of "integrated joint operations," suggesting the centrality of seizing and dominating the electromagnetic spectrum in PLA campaign theory.

Toward a Comprehensive View of Warfare

Over the past two decades, PRC civilian and military strategists have debated the nature of modern warfare. These debates draw on sources within the PLA strategic tradition and its historical experiences to provide perspective on the "revolution in military affairs," "asymmetric warfare," and "informatized" war. Such debates highlight China's interest in non-kinetic means of warfare and the increased role of economic, financial, information, legal, and psychological instruments in PLA theory and war planning. Underscoring a comprehensive, multi-dimensional view of warfare, the PLA Academy of Military Science text, *Science of Military Strategy,* notes that "war is not only a military struggle, but also a comprehensive contest on fronts of politics, economy, diplomacy, and law."

In 2003, the CCP Central Committee and the CMC approved the concept of "Three Warfares" a PLA information warfare concept aimed at influencing the psychological dimensions of military activity:

a) Psychological Warfare seeks to undermine an enemy's ability to conduct combat operations through psychological operations aimed at deterring, shocking, and demoralizing enemy military personnel and supporting civilian populations.

b) Media Warfare is aimed at influencing domestic and international public opinion to build public and international support for China's military actions and to dissuade an adversary from pursuing policies perceived to be adverse to China's interests.

c) Legal Warfare uses international and domestic laws to gain international support and manage possible political repercussions of China's military actions.

The concept of the "Three Warfares" is being developed for use in conjunction with other military and non-military operations. For example, China has incorporated the concept of Legal Warfare into its attempts to shape international opinion and interpretation of the UN Convention on the Law of the Sea away from long-accepted norms of freedom of navigation and territorial limits toward increased sovereign authority out to the 200 nautical mile Exclusive Economic Zone, the airspace above it, and possibly outer space.

Evolution of Joint Operations

A primary objective of China's military modernization programme is to develop the capacity for "Integrated Joint Operations." Launched initially in 1985, the evolution toward joint operations has been marked by three distinct conceptual phases: Collaborative Joint Operations, Coordinated Joint Operations, and Integrated Joint Operations.

- ***Collaborative Joint Operations:*** Sparse information is available on the PLA's first iteration of joint operations. It likely featured service elements acting independently, in different battle spaces, with different objectives, whose results were only indirectly mutually beneficial.

- ***Coordinated Joint Operations:*** Coordinated Joint Operations entails two or more theater service elements operating in an independent but coordinated and mutually-supporting manner toward a common objective under a unified theater commander. The PLA publication *Science of Military Campaigns* (May 2000) suggests that at the time of publication, this model had been adopted into formal instruction blocks at PLA professional military education institutions.

- ***Integrated Joint Operations:*** The content of Integrated Joint Operations has yet to be formally defined, remains largely an aspiration, and will likely continue to be so until at least 2010. Integrated

Joint Operations appears to include integration of all service elements through a common network under the leadership of a unified commander.

Obstacles: China's military leaders recognize and acknowledge that one of the primary obstacles to Integrated Joint Operations is that many PLA commanders have little or no training for, or experience operating in, a joint environment. An April 2008 directive from the CMC stressed the development of competent commanders for joint operations as the "...major priority among priorities." Other problem areas identified by the PLA include: command staff, organizing joint tactical corps, regional cooperative training, multi-service coordination training, civil-military and reserve/militia integration, insufficient preparation of small units to perform independent tasks, and lower-echelon leadership initiative.

Efforts to Improve: To rectify these deficiencies, the PLA launched enhanced training and professional military education, war simulations, and exercises. Since September 2007, the PLA has conducted at least 18 joint exercises including SHARP SWORD 2007, JOINT-2007, WARRIOR-2007, and WARRIOR- 2008. Particular areas of focus included intelligence acquisition, joint command and control, and joint logistics. In 2008 alone, PLA publications described at least four integrated joint operations exercises.

Beyond the scope of training and exercises, the PLA's joint force development may have improved due to experience gained from PLA operations in 2008 to support internal disaster relief missions, including the winter storms of February 2008 and the May 2008 Sichuan earthquake. Both of these missions may have involved the establishment of joint command posts and inter-service cooperation and coordination for extended periods of time under real world conditions.

Secrecy and Deception in PLA Military Strategy

PRC military writings point to a working definition of strategic deception as "[luring] the other side into developing misperceptions ... and [establishing for oneself] a strategically advantageous position by producing various kinds of false phenomena in an organized and planned manner with the smallest cost in manpower and materials." In addition to information operations and conventional camouflage, concealment, and denial, the PLA draws from China's historical experience and the traditional role that stratagem and deception have played in Chinese doctrine. In recent decades, there has been a resurgence of the study of classic Chinese military figures Sun Zi, Sun Bin, Wu Qi, and Shang Yang and their writings, all of which highlight the centrality of deception.

There is a contradiction between the tendencies of China's military establishment, which favours excessive secrecy, and the civilians' stated goal of reassuring neighbours and existing powers about the peaceful nature of China's development of greater influence on the world stage. This is especially true in an integrated global economy, which depends upon transparency and the free flow of information for success. The CCP's own institutional emphasis on secrecy could also lead to miscalculation or misunderstanding by outsiders of China's strategic intentions. Conversely, overconfidence among China's leaders in the uncertain and unproven benefits of stratagem and deception might lead to their own miscalculation in crises. In addition, the same skills commanders use against adversaries might also be used to slow - or cover up - the revelation of bad news internal to the PLA. Excessive reliance on secrecy and/or deception, therefore, may serve to confuse China's leaders as much foreigners about China's capabilities, doctrine, and strategic environment.

Asymmetric Warfighting

Since the 1991 Persian Gulf War and Operation Allied Force (1999) PLA military strategists have underscored the urgency of building forcestructure, strategies and tactics around new or unexpected capabilities. They also have emphasized developing innovative strategies and tactics to employ with existing technologies and weapon systems in order to level the playing field against technologically superior opponents. An article published in the *Liberation Army Daily* in 1999 posits:

"[A] strong enemy with absolute superiority is certainly not without weakness.... [our] military preparations need to be more directly aimed at finding tactics to exploit the weaknesses of a strong enemy."

Elements of China's approach to asymmetric warfare can be seen in its heavy investment in ballistic and cruise missile systems; undersea warfare systems, including submarines and advanced naval mines; counterspace systems; computer network operations; special operations forces; and the "Three Warfares" concept.

PRC Debates on Future Military Strategy

China's economic growth and development is increasingly dependent upon sustained international stability and secure access to foreign markets and resources. In recognition of these dependencies, PRC military and civilian strategists have begun to discuss the role of the armed forces in protecting and advancing

China's broader political and economic interests. The extent to which these considerations shape China's current defence policy and force planning is not known. However, it is increasingly apparent that these concerns influence China's thinking about defence planning at the highest levels.

In a speech at the March 2008 National People's Congress, CMC Vice Chairman General Xu Caihou stated that "[China's] armed forces shoulder great responsibility for safeguarding the state's sovereignty, security, territorial integrity, and for building a well-off society in an all-around way."

China's 2006 Defence White Paper states explicitly in its description of the security environment that "security issues related to energy, resources, finance, information and international shipping routes are mounting." Similarly, the 2008 Defence White Paper asserts that "struggles for strategic resources, strategic locations, and strategic dominance have intensified." It defines as among the PLA's primary tasks, "upholding national security and unity, and ensuring the interests of national development." Within the armed forces themselves, military strategists have begun shifting the focus of ground force operations from regional defence toward "trans-regional mobility," which would entail constructing task-organized mobile combat units, improved air-ground integrated operations, long-distance manoeuvre, strategic lift, expeditionary logistics, rapid assaults, and special operations. Increased PLA participation in overseas peacekeeping operations, humanitarian assistance, and disaster relief also suggests that the ground forces may become more involved in limited expeditionary operations in future.

The PLA appears to be examining how to translate these new missions and tasks into doctrine, resource allocation, force structure changes, and contingency planning. For example, there appears to be an emerging debate within the Navy over future naval strategy.

PRC President Hu Jintao called China a "sea power" and advocated a "powerful people's navy" to "uphold our maritime rights and interests" during a speech at a Navy CCP Congress in 2006. Other PRC leaders, PLA Navy officials, government writings, and PLA journals have argued that China's economic and political power is contingent upon access to and use of the sea, and that a strong navy is required to safeguard such access. Some PLA Navy thinkers, perhaps reflecting these arguments, have posited an alternative to the "Offshore Active Defence," which emphasizes coastal defence operations within the first island chain and a focus on Taiwan contingencies. The new concept, called the "Far Sea Defence," emphasizes multi-dimensional precision attacks beyond the first island chain and operations outside of China's claimed 200 nautical mile EEZ to defend PRC national interests, adding a layer of strategic depth within which to defend China's coastline.

Proponents of this strategy assert that China requires a greater number of large- and medium-size warships, carrier-based aviation, improved C4ISR, and more long-range support vessels. The PLA Navy is considering building multiple aircraft carriers and associated ships by 2020. However, some Chinese scholars have raised concerns regarding this approach, arguing that it may lead to confrontation with other nations. Regardless of whether or not China decides to pursue Far Sea Defence, China's thinking appears to be gradually moving toward a strategic concept that considers defence of maritime interests, in addition to defence of homeland, as drivers for force modernization.

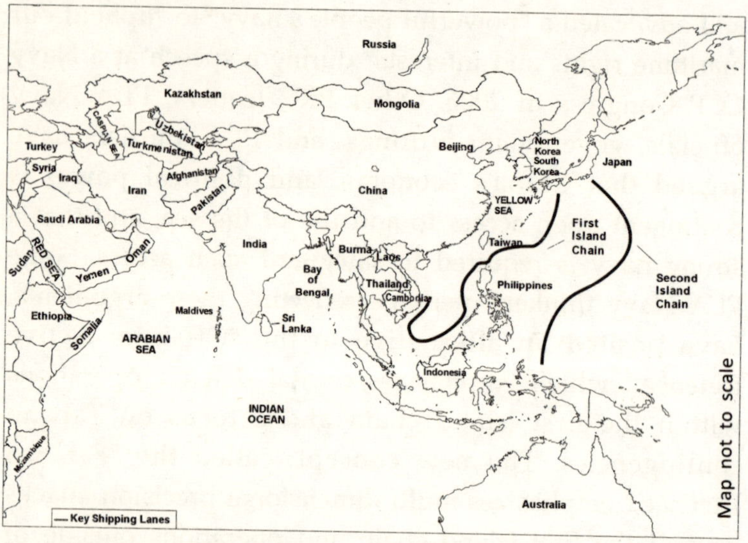

The First and Second Island Chains. ***PRC military theorists conceive of two island "chains" as forming a geographic basis for China's maritime defensive perimeter.***

The PLA's new missions are also driving internal discussions about the future of the PLA Air Force (PLAAF). An article from July 2008 argues that "in the wake of constant extension of our national interests, the change in our peripheral security environment and the struggle in air and space is getting more acute, the strategic needs that national interests impose on air security are also increasing." The article highlighted new or expanded missions that will affect the PLAAF in the future:

a) Control the airspace over China's territory, maintain air surveillance over China's maritime interests, defend against "foreign harassing attacks from space," and protect China's national dignity, sovereignty, rights, and interests.

b) Provide air security for China's overseas investment, communications and transport, scientific survey, and other efforts related to China's economic modernization.

c) Prepare for Taiwan contingencies; support operations against separatists, terrorists, and other criminal forces; maintain China's territorial integrity and social stability; and protect Chinese citizens and property.

d) Defeat enemy's air capabilities, ensure China's air defence, and contribute to integrated joint operations during wartime.

CHAPTER 3

Force Modernization: Goals and Trends

"[The PLA] has formulated in a scientific way strategic plans for national defence and armed forces building and strategies for the development of the services and arms, according to which it will lay a solid foundation by 2010, basically accomplish mechanization, make major progress in informatization by 2020, and by and large reach the goal of modernization of national defence and armed forces by the mid-21st Century."

- China's National Defence in 2008

Overview

China's leaders have stated their intentions and allocated resources to pursue broad-based military transformation that encompasses force-wide professionalization; improved training; more robust, realistic joint exercises; and accelerated acquisition and development of modern conventional and nuclear weapons. China's military appears focused on assuring the capability to prevent Taiwan independence and, if Beijing were to decide to adopt such an approach to compel the island to negotiate a settlement on Beijing's terms. At the same time, China is laying the foundation for a force able to accomplish broader regional and global objectives.

The U.S. Intelligence Community estimates China will take until the end of this decade or longer to produce a modern force capable of defeating a moderate-size adversary. China will not be able to project and sustain small military units far beyond China before 2015, and will not be able to project and sustain large forces in combat operations far from China until well into the following decade. The PLA continues to face deficiencies in inter-service cooperation and actual experience in joint exercises and combat operations. Recognizing these shortcomings, China's leaders stress asymmetric strategies to leverage China's advantages while exploiting the perceived vulnerabilities of potential opponents using so-called "Assassin's Mace" programmes (e.g., counterspace and cyberwarfare programmes).

Anti-Access/Area-Denial Capability Developments

As part of its planning for a Taiwan contingency, China is prioritizing the development of measures to deter or counter third-party intervention in any future cross-Strait crisis. China's approach to dealing with this challenge appears to be reflected in a sustained effort to develop the capability to attack, at long ranges, military forces that might deploy (anti-access) or operate (area-denial) within the western Pacific. In this context, China's anti-access/area-denial forces increasingly overlap, providing multiple layers of offensive systems utilizing the sea, air, space, and cyber-space.

Potential for Miscalculation

As the PLA modernizes, three misperceptions could lead to miscalculation or crisis. First, other countries could underestimate the extent to which PLA forces have improved. Second, China's leaders could overestimate the proficiency of their forces by assuming new systems are fully operational, adeptly operated, adequately maintained, and well integrated with existing or other new capabilities. Third, China's leaders may fail to appreciate the effects of their decisions on the security perceptions and responses of other regional actors.

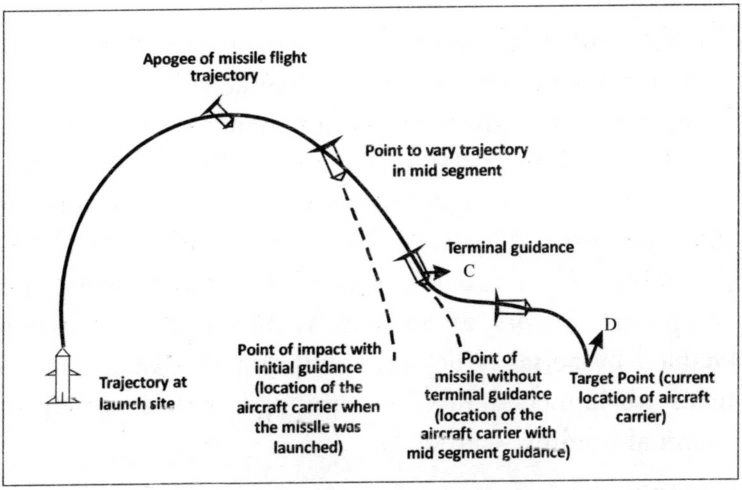

Schematic Diagram of Missile Flight Trajectory with Terminal Guidance. *This is a graphic of the Anti-Ship Ballistic Missile's use of mid-course and terminal guidance to strike an aircraft carrier used in a 2006 article from the Second Artillery Engineering College.*

Analyses of current and projected force structure improvements suggest that China is seeking the capacity to hold surface ships at risk through a layered capability reaching out to the "second island chain." One area of investment involves combining conventionally-armed anti-ship ballistic missiles (ASBMs) based on the CSS-5 (DF-

21) airframe, C4ISR for geo-location and tracking of targets, and onboard guidance systems for terminal homing to strike surface ships. As described in an authoritative 2004 article for the Second Artillery Corps, the ASBM could employ "terminal-sensitive penetrating sub-munitions" to "destroy the enemy's carrier-borne planes, the control tower and other easily damaged and vital positions." This capability would have particular significance, as it would provide China with preemptive and coercive options in a regional crisis.

PRC military analysts have also concluded that logistics and mobilization are potential vulnerabilities in modern warfare, given the requirements for precision in coordinating transportation, communications, and logistics networks. To threaten regional bases, logistics, and support infrastructure, China could employ SRBM/MRBMs, ground-launched LACMs, special operations forces, and computer network attack (CNA). Strike aircraft, when enabled by aerial refueling, could engage distant targets using air-launched LACMs equipped with a variety of terminal-homing warheads.

China's emerging local sea-denial capabilities - mines, submarines, maritime strike aircraft, and modern surface combatants equipped with advanced ASCMs - provide a supporting layer of defence for its long-range anti-access systems. Acquisition and development of the KILO, SONG, SHANG, and YUAN-class submarines illustrates the importance the PLA places on undersea warfare for sea denial. In the past ten years, China has deployed ten new classes of ships. The purchase of SOVREMENNYY II-class DDGs and indigenous production of the LUYANG I/

LUYANG II DDGs equipped with long-range ASCM and surface-to-air missile (SAM) systems, for example, demonstrate a continuing emphasis on improving anti-surface warfare, combined with mobile, wide-area air control.

The air and air defence component of anti-access/area-denial includes SAMs such as the HQ-9, SA-10, SA-20 (which has a reported limited ballistic and cruise missile defence capability), and the extended-range SA-20 PMU2. Beijing will also use Russian-built and domestic fourth-generation aircraft (e.g., Su-27 and Su-30 variants, and the indigenous F-10 multi-role fighter) to compete for local air dominance. The PLA Navy would employ Russian Su-30MK2 fighters, armed with AS-17/Kh-31A anti-ship missiles, and FB-7 fighter-bombers for maritime interdiction. Acquisition of an air refueling platform like the Russian IL-78 would extend operational ranges for PLAAF and PLA Navy strike aircraft armed with precision munitions, thereby increasing the threat to surface and air forces, bases, and logistics nodes distant from China's coast. Additionally, acquisition and development of longer-range unmanned aerial vehicles (UAVs) and UCAVs, including the Israeli HARPY, expands China's options for long-range reconnaissance and strike.

A final element of an emerging area anti-access/anti-denial strategy includes the electromagnetic and information spheres. PLA authors often cite the need in modern warfare to control information, sometimes termed "information blockade" or "information dominance," and to seize the initiative as a critical function in the early phases of a campaign. China is improving information and operational security. It is also developing electronic and information warfare capabilities, as well as denial and deception strategies. China's "information blockade" likely envisions employment of military and non-military instruments of state power across all dimensions of the modern battlespace, including outer space.

Building Capacity for Conventional Precision Strike

Short-Range Ballistic Missiles (SRBMs) (< 1,000 km): As of September 2008 the PLA had 1,050-1,150 SRBMs and is increasing its inventory at a rate of over 100 missiles per year. The PLA's first-generation SRBMs do not possess true "precision strike" capability; later generations have greater ranges, improved accuracy, and a wider variety of conventional payloads, including unitary and submunition warheads.

Medium-Range Ballistic Missiles (MRBMs) (1,000-3,000 km): The PLA is acquiring conventional MRBMs to increase the range to which it can conduct precision strikes, to include targeting naval ships, including aircraft carriers, operating far from China's shores.

Land-Attack Cruise Missiles (LACMs): The PLA is developing air- and ground-launched LACMs, such as the YJ-63 and DH-10 systems for stand-off, precision strikes. As of April 2008 the PLA had 150-350 DH-10 ground-launched cruise missiles.

Air-to-Surface Missiles (ASMs): According to DIA estimates, the PLA has a small number of tactical ASMs and precision-guided munitions, including all-weather, satellite- and laser-guided bombs, and is pursuing improved airborne anti-ship capabilities.

Anti-Ship Cruise Missiles (ASCMs): The PLA Navy has or is acquiring nearly a dozen ASCM variants, ranging from the 1950s-era CSS-N-2 to the modern Russian-made SS-N-22 and SS-N-27B. The pace of ASCM research, development and production - and of foreign procurement - has accelerated over the past decade.

Anti-Radiation Weapons: The PLA has imported Israeli-made HARPY unmanned combat aerial vehicles (UCAVs) and Russian-made anti-radiation missiles (ARM), and is developing an ARM based on the Russian Kh-31P (AS-17) known as the YJ-91.

Artillery-Delivered High Precision Munition: The PLA is deploying the A-100 300 mm multiple rocket launcher (MRL) (100+ km range) and developing the WS-2 400 mm MRL (200 km range).

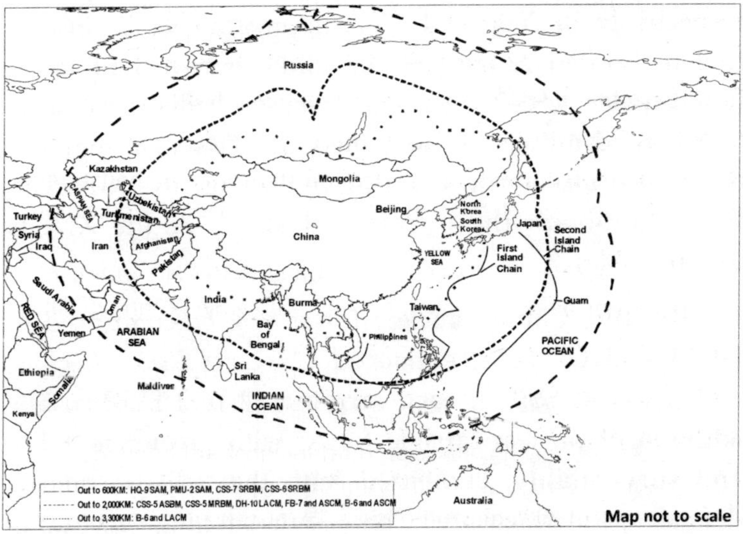

Conventional Anti-Access Capabilities. *The PLA's conventional forces are currently capable of striking targets well beyond China's immediate periphery. Not included are ranges for naval surface- and sub-surface-based weapons, whose employment at distances from China would be determined by doctrine and the scenario in which they are employed.*

Strategic Capabilities

China has made steady progress in recent years in developing offensive nuclear, space, and cyber warfare capabilities - the only aspects of China's armed forces that, today, have the potential to be truly global. However, there is little evidence that China's military and civilian leaders have fully thought through the global and systemic effects of the use of these capabilities.

Nuclear Forces: China is both qualitatively and quantitatively improving its strategic missile forces. China's nuclear arsenal currently consists of approximately 20 silo-based, liquid-fueled CSS-4 ICBMs; solid-fueled, road-mobile DF-31 and DF-31A ICBMs, which were deployed respectively in 2006 and 2007; approximately 20 liquid-fueled, limited range CSS-3 ICBMs; between 15 to 20 liquid-fueled CSS-2 intermediate-range ballistic missiles; CSS-5 road-mobile, solid-fueled MRBMs (for regional deterrence missions); and JL-1 SLBMs on the XIA-class SSBN (although the operational status of the XIA is questionable).

By 2010, China's nuclear forces will be comprised of DF-31 and DF-31As; enhanced CSS-4s; CSS-3s; CSS-5s; and JIN-class SSBNs, each carrying 12 JL-2 SLBMs. The addition of nuclear-capable forces with greater mobility and survivability, combined with the ballistic missile defence countermeasures China is researching - including manoeuvring re-entry vehicles (MaRV), multiple independently targeted re-entry vehicles (MIRV), decoys, chaff, jamming, thermal shielding, and ASAT weapons - will strengthen China's deterrent and enhance its strategic strike capabilities. New air- and ground-launched cruise missiles that could potentially perform nuclear missions would similarly improve the survivability, flexibility, and effectiveness of China's nuclear forces.

The introduction of more mobile systems will create new command and control challenges for China's leadership, which now confronts a different set of variables related to release and deployment authorities. For example, the PLA has only a limited capacity to communicate with submarines at sea and the PLA Navy has no experience in managing a SSBN fleet that performs strategic patrols. Of note, recent missile force training, as described in China's state-owned press, has included scenarios in which missile batteries lose communication links with higher echelons and other situations requires commanders to choose alternative launch locations. This training may provide limited insights into how the PLA may be seeking to address this issue.

China's 2008 Defence White Paper states that the Second Artillery Corps: 1) sticks to China's policy of no first-use of nuclear weapons, 2) implements a self-defensive nuclear strategy, 3) strictly follows the orders of the CMC, and 4) takes as its fundamental mission the protection of China from any nuclear attack. The 2008 Defence White Paper also states that "if China comes under a nuclear attack, the nuclear missile force of the Second Artillery Force [sic] will use nuclear missiles to launch a resolute counterattack against the enemy either independently or together with the *nuclear forces of other services*" [emphasis added]. This implies nuclear missions may be assigned to the PLAAF, in addition to the PLA Navy. PRC military writings suggest that additional missions for China's nuclear forces include deterring conventional attacks against PRC nuclear assets or conventional attacks with WMD-like effects, reinforcing China's great power status and increasing its freedom of action by limiting the extent to which others can coerce China with nuclear threats.

Given the above missions for China's nuclear forces, the conditions under which China's "no first use" policy applies are unclear. The PRC government has provided public and private assurances that its "no first use" policy has not and will not change. PRC writings indicate internal PLA support for this policy. Nevertheless, periodic PRC military and civilian academic debates have occurred over whether a "no first use" policy supports or detracts from China's deterrent, and whether or not "no first use" should remain in place. Questions also continue regarding whether or not a conventional strike on China's strategic forces would nullify China's "no first use" pledge. These debates add a further layer of ambiguity to China's strategic intentions for its nuclear forces.

Space and Counterspace: China's space activities and capabilities, including ASAT programmes, have significant implications for anti-access/area-denial in Taiwan Strait contingencies and beyond. Many of China's space programs, including the manned programme and the planned space station, are run by the PLA. China views the development of space and counterspace capabilities as bolstering national prestige and, like nuclear weapons, demonstrating the attributes of a great power.

Reconnaissance: China is deploying advanced imagery, reconnaissance, and Earth resource systems with military applications. Examples include the Yaogan-1, -2, -3, -4, and -5, the Haiyang-1B, the CBERS-2 and -2B satellites, and the Huanjing disaster/environmental monitoring satellite constellation. China is planning eight satellites in the Huanjing programme that are capable of visible, infrared, multi-spectral, and synthetic aperture radar imaging. In the next decade, as Beijing fields a more robust constellation of reconnaissance satellites, it probably will employ commercial satellite imagery to supplement existing coverage.

Navigation and Timing: China is pursuing multiple possibilities for satellite navigation independence. Currently, the PRC uses the U.S. global positioning system (GPS), Russia's GLONASS, and its own BeiDou-1 (regional) systems for navigation. The BeiDou-1 system consists of three satellites and serves both civil and military purposes. The Beidou-1 system will be replaced by a BeiDou-2 system (expected to be operational in 2011) that will become a regional complement to the worldwide BeiDou-2/Compass system expected to be operational in 2015-2020.

Manned Space and Lunar Programmes: China successfully performed its first space walk in September 2008 from the Shenzhou-VII, which was preceded by the October 2007 launch of its first lunar orbiter, the Chang'e-1. China's goals are to have a manned space station and to conduct an unmanned lunar landing and return mission by 2020. The manned space programme probably benefits PLA weapons development programmes.

a) Rocket and control system capabilities required for the Shenzhou-VII mission may have applications for ballistic missile development.

b) During its mission, the Shenzhou-VII deployed the Banxing-1 (BX-1), a small imaging satellite, which successfully positioned itself into an orbit around the orbital module. The stated purpose of this technology is to monitor instrumentation in space and detect malfunctions. Further applications could support counterspace activities.

Communications: China uses commercial, consortium, and civil communications satellites (COMSATs) for both regional and international telecommunications, to include satellite television, internet, and telephony. Along with regional development of related technologies, China has recently entered the world market by exporting COMSATs and infrastructure to Venezuela and Nigeria. In April 2008, China launched its first data-relay satellite, the TianLian-1.

Small Satellites: Since 2000, China has launched a number of small satellites, including oceanographic research, imagery, and environmental research satellites. China has also established small satellite design and production facilities and is developing microsatellites - weighing less than 100 kilograms - for remote sensing, and networks of imagery and radar satellites. These developments could allow for a rapid reconstitution or expansion of China's satellite force in the event of any disruption in coverage, given an adequate supply of boosters. Beijing's effort to develop small, rapid-reaction space launch vehicles currently appears to be stalled.

ASAT Weapons: In January 2007, China successfully tested a direct-ascent ASAT missile against a PRC weather satellite, demonstrating its ability to attack satellites in low-Earth orbit. The direct-ascent ASAT system is one component of a multi-dimensional programme to limit or prevent the use of space-based assets by potential adversaries during times of crisis or conflict.

China's nuclear arsenal has long provided Beijing with an inherent ASAT capability. Ultra High Frequency (UHF)-band satellite communications jammers acquired from Ukraine in the late 1990s along with probable indigenous systems give China the capacity to jam common satellite communications bands and GPS receivers. In addition to the direct-ascent ASAT programme, China is developing other technologies and concepts for kinetic and directed-energy (e.g., lasers, high-powered microwave, and particle beam) weapons for ASAT missions. Citing the requirements of its manned and lunar space programmes, China is improving its ability to track and identify satellites - a prerequisite for effective, precise counterspace operations.

Information Warfare: There has been much writing on information warfare among China's military thinkers, who indicate a strong conceptual understanding of its methods and uses. For example, a November 2006 *Liberation Army Daily* commentary outlines:

"[The] mechanism to get the upper hand of the enemy in a war under conditions of informatization finds prominent expression in whether or not we are capable of using various means to obtain information and of ensuring the effective circulation of information; whether or not we are capable of making full use of the permeability, sharable property, and connection of information to realize the organic merging of materials, energy, and information to form a combined fighting strength; [and,] whether or not we are capable of applying effective means to weaken the enemy side's information superiority and lower the operational efficiency of enemy information equipment."

The PLA is investing in electronic countermeasures, defences against electronic attack (e.g., electronic and infrared decoys, angle reflectors, and false target generators), and Computer Network Operations (CNO). China's CNO concepts include computer network attack (CNA), computer network exploitation (CNE), and computer network defence (CND). The PLA has established information warfare units to develop viruses to attack enemy computer systems and networks, and tactics and measures to protect friendly computer systems and networks. In 2005, the PLA began to incorporate offensive CNO into its exercises, primarily in first strikes against enemy networks.

Advances in C4ISR

China's military strategists have identified reliable, survivable, interoperable, and integrated C4ISR systems as essential for coordinating joint operations and for engaging successfully on the modern battlefield. Establishing modern C4ISR, in short, is a central task to accomplish China's objective to build forces to fight and win "local wars under informatized conditions." Consequently, China is acquiring advanced land, air, sea, and space-based C4ISR capabilities to enhance battlefield awareness; identify, track, and engage military targets deep into the western Pacific Ocean; and, streamline supply and logistics functions.

The objective of the C4ISR network envisioned by the PLA appears to focus on theater-level integration of campaign and tactical command networks as a means to fuse communications, intelligence and reconnaissance, electronic countermeasures, and early warning systems. Through improvements in command and staff training, the PLA is seeking to increase battlefield commanders' abilities to employ automated, decision-support systems in a dynamic combat environment, including what PLA theorists term as a "complex electromagnetic environment."

Power Projection - Modernization Beyond Taiwan

China continues to invest in military programmes designed to improve extended-range power projection. Current trends in China's military capabilities are a major factor in changing East Asian military balances, and could provide China with a force capable of conducting a range of military operations in Asia well beyond Taiwan. Given the apparent absence of direct threats from other nations, the purposes to which China's current and future military power will be applied remain uncertain. These capabilities will increase Beijing's options for military coercion to press diplomatic advantage, advance interests, or resolve disputes in its favour.

Some PLA analysts have explored the geopolitical value of Taiwan in extending China's maritime "defensive" perimeter and improving its ability to influence regional sea lines of communication. For example, the PLA Academy of Military Science text Science of Military Strategy (2000) states:

"If Taiwan should be alienated from the Mainland, not only [would] our natural maritime defence system lose its depth, opening a sea gateway to outside forces, but also a large area of water territory and rich resources of ocean resources would fall into the hands of others.... [O]ur line of foreign trade and transportation which is vital to China's opening up and economic development will be exposed to the surveillance and threats of separatists and enemy forces, and China will forever be locked to the west of the first chain of islands in the West Pacific."

The related desire to protect energy investments in Central Asia and land lines of communication could also provide an incentive for military investment or intervention if instability surfaces in the region. Disagreements with Japan over maritime claims in the East China Sea and with several Southeast Asian claimants to all or parts of the Spratly and Paracel Islands in the South China Sea could lead to renewed tensions in these areas. Instability on the Korean Peninsula likewise could produce a regional crisis.

Analysis of China's weapons acquisitions also suggests China is looking beyond Taiwan as it builds its force. For example, new missile units outfitted with conventional theater-range missiles at various locations in China could be used in a variety of non-Taiwan contingencies. Airborne early warning and control (AEW&C) and aerial-refueling programmes would permit extended air operations into the South China Sea and beyond.

Advanced destroyers and submarines reflect Beijing's desire to protect and advance its maritime interests up to and beyond the second island chain. Expeditionary forces (three airborne divisions, two amphibious infantry divisions, two marine brigades, and about seven special operation groups) are improving with the introduction of new equipment, better unit-level tactics, and greater coordination of joint operations. Over the long term, improvements in China's C4ISR, including space-based and over-the-horizon sensors, could enable Beijing to identify, track, and target military activities deep into the western Pacific Ocean.

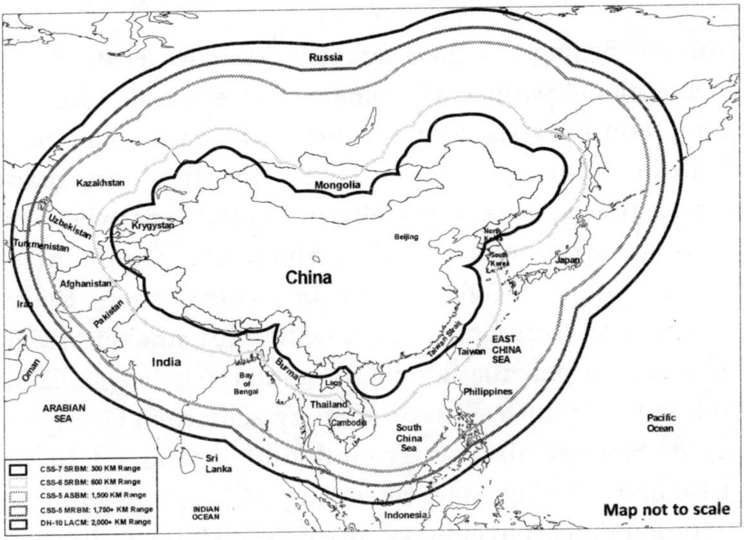

***Regional Conventional Missiles.** China is capable of employing land-based ballistic and cruise missile systems to support a variety of regional contingencies.*

India: China has deepened its ties with India through increased trade, high-level dialogues, and an improved military-to-military relationship. China and India agreed to boost trade from $11.4 billion in 2007 to $40 billion in 2010, and they have held several rounds of dialogue over disputed territorial claims. Sino-Indian defence ties were institutionalized in 2007 with the establishment of an Annual Defence Dialogue and by conducting three bilateral defence exercises since 2007. However, the PLA remains concerned with persistent disputes along China's shared border with India and the strategic ramifications of India's rising economic, political, and military power. The PLA has replaced older liquid-fueled nuclear-capable CSS-3 IRBMs with more advanced solid-fueled CSS-5 MRBMs in Western China, and may possibly be developing contingency plans to move airborne troops into the region.

Russia: Beijing continues to view Moscow as its closest international partner, yet remains concerned that Russia's long-term interests do not correspond with China's. Sino-Russia bilateral cooperation continues on many international issues, especially in Central Asia where the two jointly manage the Shanghai Cooperation Organization (SCO). Despite this cooperation, Russia resents and fears China's rise, while PLA strategists view Russia as a potential long-term military challenge. China refused to endorse Russian military activity in Georgia in 2008. Several air and ground units stationed in the Lanzhou, Beijing, and Shenyang Military Regions, in addition to its conventional missile and strategic deterrent forces, could be used in Russia contingencies.

Central Asia: China's primary interests in Central Asia are centered on building regional influence, obtaining natural resources and energy, and countering support for China's Uighur separatists. Beijing has reached agreements with many Central Asian governments to build the infrastructure necessary to transport resources into western China, such as a pipeline that will stretch from Turkmenistan through Uzbekistan and Kazakhstan into China. Beijing has also conducted bilateral and multilateral exercises with SCO member states to enhance China's influence within the SCO and to build cohesive regional opposition to Uighur activities. Internal security forces in Xinjiang could be used in Central Asian contingencies, and army aviation and trans-regional mobility operations could be applied to rapidly deploy combat power to the region in a crisis.

CHAPTER 4

Resources for Force Modernization

"We need to build an innovative system of defence science and technology ... that integrates military and civilian scientific-technological resources, and that organically integrates basic research, applied R&D, product designing and manufacturing, and procurement to technologies and products so as to create a good structure under which military and civilian high technologies are shared and mutually transferable."

- PRC President Hu Jintao

Overview

Sources for PLA modernization include domestic defence expenditures, indigenous defence industrial development, dual-use technologies, and foreign technology acquisition - all of which are driven by the performance of the economy. As China's defence industries develop, the PLA is relying on acquisition of foreign weapons and technology, primarily from Russia, to fill near-term capability gaps. China also harvests spin-offs from foreign direct investment and joint ventures in the civilian sector, technical knowledge and expertise of students returned from abroad, and state-sponsored industrial espionage to increase the level of technologies available to support military research, development, and acquisition. Beijing's long-term goal is to create a wholly indigenous defence industrial sector able to meet the needs of PLA modernization and to compete as a top-tier producer in the global arms trade. China is already competitive in some areas, such as communications, with leading international defence firms.

Military Expenditure Trends

On March 4, 2008, Beijing announced a 17.6 percent increase in its military budget to approximately $60 billion. China's military budget doubled between 1989 and 1994, and almost doubled again between 1994 and 1999. The 2005 military budget was almost ten times the 1989 military budget. If these trends continue, China's military budget for 2009 will nearly double the 2005 figure.

China continues a two-decade trend of double digit percentage annual increases in its military budget. These increases surpass the percentage increases of its overall economic growth. Analysis of 1996-2008 data indicates that China's officially disclosed defence budget grew at an average of 12.9 percent in real terms over the period, while GDP grew at 9.6 percent.

It is unclear if China's military budget will be supplemented to recoup costs incurred during the 2008 snow storm operations, domestic suppression operations in Tibet, rescue, recovery and rebuilding after the Sichuan earthquake, and for security operations supporting the

2008 Summer Olympics. It is also unclear if, or how, the international financial downturn will affect the relative priority China's leaders will give to the military budget in the face of possible declines in China's economy and the resulting reduction in the government's tax revenue.

Estimating China's Actual Military Expenditures: The Department of Defence estimates China's total military-related spending for 2008 to be between $105 billion and $150 billion, using 2007 prices and exchange rates.

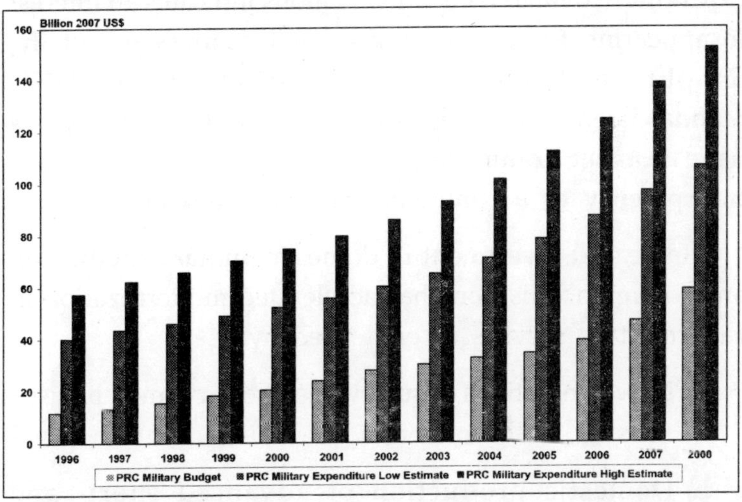

PRC Military Budget and Estimated Expenditures, 1996 - 2008. *The graphic depicts China's official military budget since 1996 and associated DoD estimates of actual military expenditures. DoD estimates include projected expenses for strategic forces, foreign acquisitions, military R&D, and paramilitary forces. All figures are in 2007 U.S. dollars.*

Estimating actual PLA military expenditures is a difficult process due to the lack of accounting transparency and China's still incomplete transition from a command economy. Moreover, China's published military budget does not include major categories of expenditure. China's legislature does not have an oversight process for the PLA budget. Although academic experts and outside analysts may disagree about the exact amount of military expenditure in China, almost all arrive at the same conclusion: Beijing significantly under-reports its military expenditures.

The United States and other countries have for many years urged China to increase transparency in military spending. In September 2008, China submitted an annual report on its military expenditure to the UN Secretary General. While the decision to resume reporting of military expenditure data to the United Nations indicates an interest in appearing to be more transparent, China's use of the Simplified Reporting Form instead of the UN's Standardized Reporting Form suggests China's leaders have not yet committed fully to the idea of military transparency as a confidence-building measure.

Increased investment in domestic military production and foreign acquisitions has accelerated modernization in each military service, as evidenced by:

a) New generations of survivable nuclear armed ballistic missiles, both land- and sea-based;

b) Domestic production of advanced short- and medium range conventionally armed ballistic missiles;

c) Advanced attack and ballistic missile submarines and associated weaponry;

d) Advanced Russian aircraft and precision weaponry for the air and naval air forces;

e) Domestic development of the multi-role F-10 fighter aircraft;

f) Advanced Russian and domestic versions of modern guided-missile destroyers (DDGs), frigates, and amphibious landing craft;

g) Modern, long-range, and mobile air defence systems; and,

h) Programmes to increase professionalism and quality of life for military personnel.

PEOPLE'S REPUBLIC OF CHINA (millions of US$) FISCAL YEAR: 2007

	Active Forces	Reserve Forces	Militia	Total: Amount	Total: Percentage (%)
Personnel	15,638	137	0	15,776	33.76
Training & Maintenance	14,757	199	954	15,911	34.05
Equipment	14,748	149	145	15,042	32.19
Total	**45,144**	**485**	**1,099**	**46,729**	**100**

Notes:
1. Personnel expenses cover salaries, allowances, food, clothing and bedding, insurance, welfare benefits and pensions for officers, non-ranking cadres, enlisted men, and contracted civilians.
2. Training and maintenance expenses cover troop training, institutional education, and running and development of daily work and activities.
3. Equipment expenses cover R&D, procurement, maintenance, and transportation and storage of weaponry and equipment.

***PRC Submission to United Nations on Military Expenditures, 2007.** In September 2008, China submitted a report to the United Nations on defence expenditures. While it provides some new insight into funding for active, reserve, and militia forces, China's use of the Simplified Reporting Form indicates Beijing is not yet fully committed to the idea of transparency as a confidence-building measure.*

China's Advancing Defence Industries

Since the late 1990s, China's state-owned defence and defence-related companies have undergone broad-based transformation. Beijing is improving business practices, streamlining bureaucracy, shortening development timelines, boosting quality control, and increasing production capacity for military orders. Beijing is also emphasizing integration of defence and non-defence sectors to leverage the latest dual-use technologies and the output from China's expanding science and technology base. Augmented by direct acquisition of foreign weapons and technology, these reforms have enabled China to develop and produce advanced weapon systems such as missiles, fighter aircraft, and warships.

Increasing Efficiency and Capacity: China's 2006 Defence White Paper notes that the output value, added value, and gross revenue of defence-related science, technology, and industry increased by 24.3 percent, 20.7 percent, and 21.6 percent, respectively, in 2005. A similar analysis was not included in the 2008 Defence White Paper.

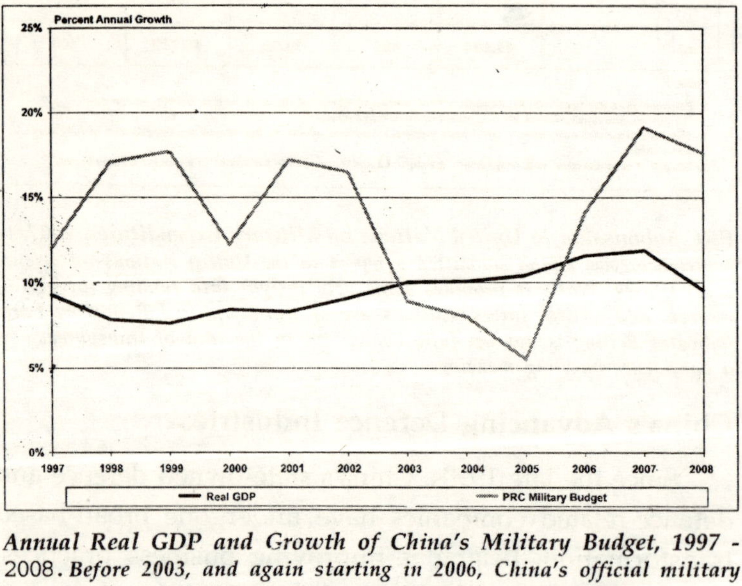

Annual Real GDP and Growth of China's Military Budget, 1997 - 2008. *Before 2003, and again starting in 2006, China's official military budget grew at a higher percentage rate than China's real GDP.*

Through at least the 11th Five-Year Plan period (2006-2010), China's defence-related industries will continue to reap benefits from:

a) Transfers of technology and skills from foreign joint ventures;

b) Increased government funding for research, development, and procurement;

c) The manned space flight programme, including its vessels and tracking stations;

d) Legal and illegal acquisition of foreign military and dual-use technology;

e) Increased partnerships with academic institutions, which improve student recruitment and technical training for existing staff; and,

f) Overseas training and experience gained by an increasing number of scientists, engineers, and managers returning to China.

Civil-Military Integration: Development of innovative dual-use technology and an industrial base that serves both military and civilian needs is among the highest priorities of China's leadership. PRC President Hu Jintao expressed in his political report to the CCP's 17th Party Congress:

"We must establish sound systems of weapons and equipment research and manufacturing ... and combine military efforts with civilian support, build the armed forces through diligence and thrift, and blaze a path of development with Chinese characteristics featuring military and civilian integration".

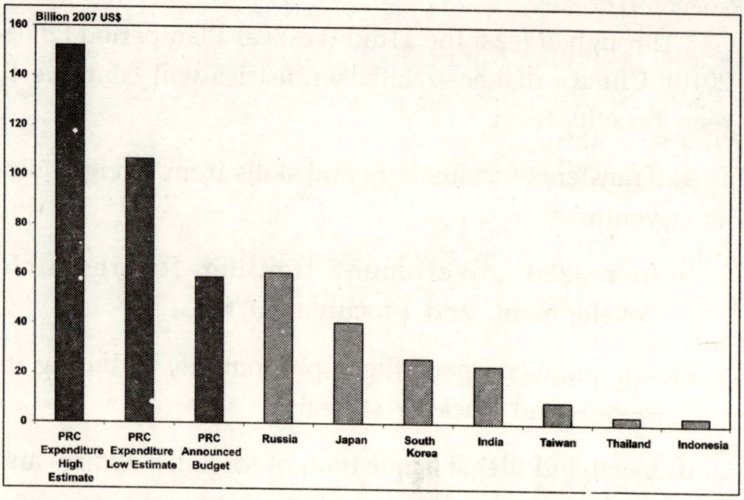

Regional Military Spending, 2008. *This graphic compares China's official and DoD-estimated military expenditures with that of other regional militaries. Military expenditures are derived from government sources. Of note, Russia's budget data grew significantly over the period of review between the 2008 and 2009 editions of this report. All figures are in 2007 U.S. dollars.*

China's defence industry has benefited from integration with China's rapidly expanding civilian economy and science and technology sector, particularly elements that have access to foreign technology. According to the Organization of Economic Cooperation and Development (OECD), China's research and development (R&D) spending has increased at an annual rate of 19 percent since 1995 to reach $30 billion in 2005, the sixth highest in the world. The OECD assessed that while China has significantly invested in R&D, human resources, and R&D infrastructure, China still has "a long way to go" to build a mature national innovation system. Progress within individual defence sectors appears to be linked to the relative integration of each - through China's civilian economy - into the global production and R&D chain. For example, the shipbuilding and defence electronics sectors, benefiting from China's leading role in producing commercial shipping and information technologies, have witnessed the greatest progress over the last decade. Information technology companies, including Huawei, Datang, and Zhongxing, maintain close ties to the PLA and collaborate on R&D. Commercial off-the-shelf technologies, such as computer network switches and routers, increasingly provide the PLA with state-of-the-art telecommunications equipment.

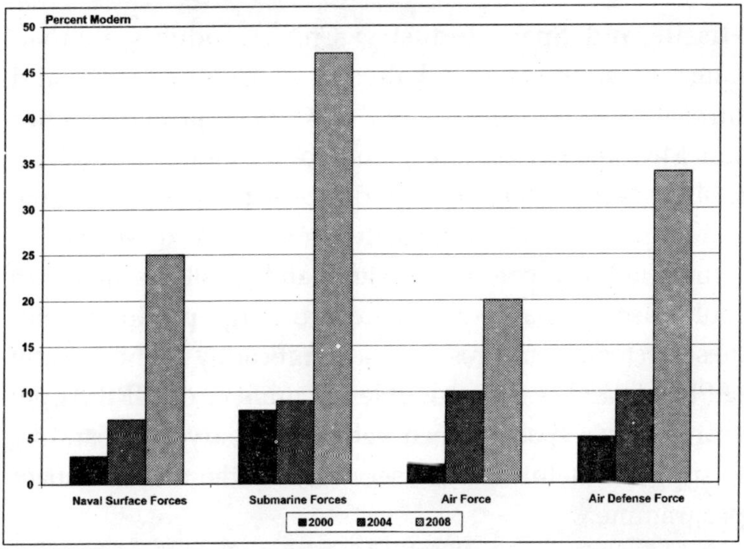

***Select PLA Modernization Areas, 2000 – 2008.** This graphic compares the expansion of modern operational systems within the PLA in 2000, 2004, and 2008.*

In contrast, enterprises producing high-performance computers, advanced applications software, and specialized top-end semiconductors/microprocessors - essential to defence microelectronics, but with limited or no counterpart in the PRC civil-industrial sector - have experienced slower progress. The aviation and ordnance sectors have similarly suffered from a lack of spin-on benefits in some critical areas, despite partnerships between foreign multinational corporations and domestic industry.

Sector-by-Sector Analysis: Progress across China's defence industry sectors has been uneven. Production trends and resource allocation appear to favour missile and space systems, followed by maritime assets (both surface and sub-surface), aircraft, and ground force materiel. In all areas, however, China is increasing the quality of its output and surge production capabilities.

Missile and Space Industry: China produces a broad range of sophisticated ballistic, cruise, air-to-air, and surface-to-air missiles. Many of China's primary SRBM and MRBM final assembly and rocket motor production facilities have received upgrades over the past few years, increasing production capacity. In addition to supplying China's military, complete systems and missile technologies could also be marketed for export. Surge production for these systems could result in a significantly higher output of SRBMs and perhaps double the number of MRBMs per year. China's space launch vehicle industry is expanding to support satellite launch services and the manned space programme.

Shipbuilding Industry: China operates a vibrant and globally competitive shipbuilding industry. China is the 2nd largest shipbuilder in the world, having surpassed Japan in 2008. Shipyard expansion and modernization have increased China's shipbuilding capacity and capability, generating benefits for all types of military projects, including: submarines; surface combatants; naval aviation, including initiatives for aircraft carriers; and amphibious/sealift-airlift assets. China continues to rely on foreign suppliers for some propulsion units and, to a lesser degree, fire control systems, cruise missiles, ship-to-air missiles, torpedo systems, sensors, and other advanced electronics. Modular shipbuilding techniques will allow China to spread production across multiple locations, increasing both efficiency and output. China has already demonstrated an ability to surge submarine and amphibious production.

Armament Industry: China's ground force modernization includes production of new tanks, armored personnel carriers, and artillery pieces. There have been advances in almost every area of PLA ground forces with new production capacity to accommodate surge requests. China's reliance on foreign partners to fill gaps in critical technical capabilities could still limit actual surge output.

Aviation Industry: China's commercial and military aviation industries have advanced from producing direct copies of early Soviet models to developing and producing indigenous aircraft. These include improved versions of older aircraft and modern fourth generation fighters. China's commercial aircraft industry has imported high-precision and technologically advanced machine tools, electronics, and other components that can also be used in the production of military aircraft. However, China's ability to surge production in the aircraft industry will be limited by its reliance on foreign sourcing for aircraft engines and avionics, as well as the availability of skilled personnel and facilities.

Foreign Technology Acquisition: Key areas where China continues to rely most heavily on foreign technologies include: guidance and control systems, turbine engine technology, and enabling technologies such as precision machine tools, advanced diagnostic and forensic equipment, applications and processes essential to rapid prototyping, and computer-assisted design/manufacturing (CAD/CAM). China often pursues these foreign technologies for the purpose of reverse engineering.

Russia in recent years has been China's primary weapons and materiel provider, selling Beijing advanced fighter aircraft, missile systems, submarines, and destroyers. Relying on Russian components for several of its production programs, China purchased production rights to Russian weapon designs and is negotiating the purchase of several advanced systems. Additionally, Russia cooperates with China on technical, design, and material support for numerous weapons and space systems.

Israel previously has supplied advanced military technology to China, but recently reformed its export control regime through the passage of a Defence Export Control Act in July 2007 and the adoption of implementing regulations in December 2007.

Since 2003, China has been pressuring the European Union (EU) Member States to lift the embargo on lethal military sales to China that the EU imposed in response to China's 1989 crackdown on Tiananmen Square demonstrators. In their Joint Statement following the 2004 EU-China Summit, European and PRC leaders committed to work towards lifting the embargo. Although the issue officially remains on the EU agenda, there is no consensus among the EU Member States on lifting the embargo any time in the near future.

China continues a systematic effort to obtain dual-use and military technologies from abroad through legal and illegal commercial transactions. Many dual-use technologies, such as software, integrated circuits, computers, electronics, semiconductors, telecommunications, and information security systems, are vital for the PLA's transformation into an information-based, network-enabled force. Several high-profile legal cases highlight China's efforts to obtain sensitive U.S. technologies (e.g., missile, imaging, semiconductor, and submarine) illegally by targeting well-placed scientists and businessmen.

Current and former senior U.S. intelligence officials continue to cite China as posing a growing threat to national security due to China's sustained efforts to obtain U.S. technology illegally. Many of the technologies identified during investigations into the illicit operations of PRC-based entities publicized within the last year involve sensitive military systems or programs, such as military source code, night vision equipment, cruise missile technology, and dual-use systems and components for radar and communications equipment. According to a 2008 Federal Bureau of Investigation (FBI) statcment, PRC intelligence services "pose a significant threat both to the national security and to the compromise of U.S. critical national assets," and concluded that these services "will remain a significant threat for a long time." The U.S. intelligence community has noted that, of all foreign intelligence organizations attempting to penetrate U.S. agencies, China's are the most aggressive.

Logistics Reform

Over the last decade, the PLA has improved its capability to support operations within its borders and along its periphery. Frequent training in mobility operations; improvements to command, control and coordination; and standardization of warehouse systems have strengthened the PLA's overall ability to mobilize and support local military operations. Integration of automated logistics systems into PLA command and control systems and civil logistics capabilities into military support systems will further improve the PLA's logistics capability.

The absence of a true expeditionary logistics capability, however, will limit the PLA's ability to project and sustain military operations at locations distant from the mainland. First among these limitations is the capability to transport and sustain more than one division of ground troops and equipment by sea or air. The PLA Navy's total amphibious lift capacity has been estimated to be one infantry division of approximately 10,000 troops and equipment at one time. Likewise, if all large transport aircraft in the PLAAF were operational and rigged for parachute drop, China could deliver about 5,000 parachutists in a single lift - less if equipment is carried at the same time. PLA in-flight refueling capability is also limited and can support only small numbers of fighter aircraft. Although the PLA Navy has gained some proficiency with underway replenishment and sustainment of long distance deployments, this capability remains limited by the PLA's small numbers of support ships.

The PLA's force projection capabilities will remain limited over the next decade as the PLA replaces outdated aircraft and maritime vessels and adjusts operational doctrine to encompass new capabilities. These changes will require tailored logistics equipment and training that will take time and funding to develop. Although foreign-produced or civil sector equipment and maintenance parts may help to fill near-term gaps, continued reliance on non-organic assets will hinder PLA capabilities to sustain large-scale operations.

Looking to the Future: Trends and Projections

China's National Medium- and Long-Term Programme for Science and Technology Development (2006-2020), issued by the State Council in February 2006, seeks to transform China into an "innovation-oriented society by 2020." The plan defines China's science and technology focus in terms of "basic research," "leading-edge technologies," "key fields and priority subjects," and "major special items" - all of which have military applications.

Basic Research: As part of a broad effort to expand basic research capabilities, China identified five areas that have military applications as major strategic needs or science research plans requiring active government involvement and funding:

a) Manufacturing in extreme environmental conditions,

b) Aeronautic and astronautic mechanics,

c) Information technology development, and

d) Nanotechnology research.

In nanotechnology, China has progressed from virtually no research or funding in 2002 to being a close second to the United States in total government investment.

Leading-edge Technologies: China is focusing on the following technologies for rapid development: :

a) **Information Technology:** Priorities include intelligent perception technologies, ad hoc networks, and virtual reality technologies;

b) **New Materials:** Priorities include smart materials and structures, high-temperature superconducting technologies, and highly efficient energy materials technologies;

c) **Advanced Manufacturing:** Priorities include extreme manufacturing technologies and intelligent service robots;

d) **Advanced Energy Technologies:** Priorities include hydrogen energy and fuel cell technologies, alternative fuels, and advanced vehicle technologies;

e) **Marine Technologies:** Priorities include three-dimensional maritime environmental monitoring technologies, fast, multi-parameter ocean floor survey technologies, and deep-sea operations technologies; and,

f) **Laser and Aerospace Technologies** are also high priorities.

Key Fields and Priority Subjects: China has identified certain industries and technology groups with potential to provide technological breakthroughs, remove technical obstacles across industries, and improve international competitiveness. Specifically, China's defence industries are pursuing advanced manufacturing, information technology, and defence technologies. Examples include radar, counter-space capabilities, secure C4ISR, smart materials, and low-observable technologies.

Major Special Items: China has also identified 16 "major special items" for which it plans to develop or expand indigenous capabilities. These include core electronic components, high-end universal chips and operating system software, very large-scale integrated circuit manufacturing, next-generation broadband wireless mobile communications, high-grade numerically controlled machine tools, large aircraft, high-resolution satellites, manned spaceflight, and lunar exploration.

Status of Aircraft Carrier Developments

China has an aircraft carrier research and design program, which includes continued renovations to the former Soviet Kuznetsov-class aircraft carrier VARYAG. Beginning in early 2006 with the release of China's Eleventh Five Year Plan, PRC-owned media reported high-level government and military official statements on China's intent to build aircraft carriers. In December 2008, China's Ministry of National Defence spokesman Senior Colonel Huang Xueping said "China has vast oceans and it is the sovereign responsibility of China's armed forces to ensure the country's maritime security and uphold the sovereignty of its coastal waters as well as its maritime rights and interests," and added that China is "seriously considering" adding an aircraft carrier to its fleet, because "the aircraft carrier is a symbol of a country's overall national strength, as well as the competitiveness of the country's naval force." This was preceded by a November 2008 statement by the Director of the Ministry of National Defence, Foreign Affairs Office, Major General Qian Lihua, that "having an aircraft carrier is the dream of any great military power," and "the question is not whether you have an aircraft carrier, but what you do with your aircraft carrier."

China continues to show interest in procuring Su-33 carrier-borne fighters from Russia even though the ex-VARYAG aircraft carrier has yet to complete refurbishment at Dalian shipyard. In October 2006, a Russian press report suggested early-stage negotiations were underway for China to purchase up to 50 such aircraft at a cost of

$2.5 billion. However, there has been no announcement of a contract for the aircraft.

The PLA Navy has reportedly decided to initiate a program to train 50 navy pilots to operate fixed-wing aircraft from an aircraft carrier. The programme was reported to be four years long and would be followed by ship-borne training involving the ex-VARYAG. Analysts in and out of government project that China will not have an operational, domestically-produced carrier and associated ships before 2015. However, changes in China's shipbuilding capability and degree of foreign assistance to the programme could alter those projections. The PLA Navy is considering building multiple carriers by 2020.

CHAPTER 5

Force Modernization and Security in the Taiwan Strait

"Anything that is conducive to the peaceful development of cross-Strait relations should be energetically promoted; anything that is detrimental to the peaceful development of cross-Strait relations must be firmly opposed."

- PRC President Hu Jintao, Speech at the Forum Marking the 30th Anniversary of the Issuance of "Message to Compatriots in Taiwan," 31 December 2008

Overview

The security situation in the Taiwan Strait is largely a function of dynamic interactions among Mainland China, Taiwan, and the United States. The PLA has developed and deployed military capability to coerce Taiwan or to attempt an invasion if necessary. PLA improvements pose new challenges to Taiwan's security, which has historically been based upon the PLA's inability to project power across the 100 NM Taiwan Strait, natural geographic advantages of island defence, Taiwan's armed forces' technological superiority, and the possibility of U.S. intervention.

Despite positive public statements about the Taiwan Strait situation from top leaders in Beijing following the election of Taiwan President Ma Ying-jeou, there have been no signs that Beijing's military dispositions opposite Taiwan have changed significantly. It remains to be seen how the PLA responds to PRC President Hu Jintao's call for confidence-building measures across the Strait.

Taiwan has taken important steps to build its war reserve stock, as well as improve its joint operations capability, crisis response and officer and non-commissioned officer (NCO) corps. These improvements have, on the whole, reinforced Taiwan's natural defensive advantages in the face of Beijing's continuous military build-up. Taiwan is also focused on creating an all-volunteer military and reducing its active military end-strength from 275,000 to 200,000 - or possibly 180,000 - personnel, while maintaining its defence budget at three percent of its GDP. Under this plan, the cost difference of a smaller force will free up resources to increase volunteer salaries and benefits.

The U.S. Government opposes unilateral changes to the status quo in the Taiwan Strait by either side, and calls for peaceful resolution of cross-Strait differences in a manner acceptable to the people on both sides. In accordance with the Taiwan Relations Act [Public Law 96-8, (1979)], the United States has helped to maintain peace, security, and stability in the Taiwan Strait by providing defence articles and services to support Taiwan's self-defence. In addition, the U.S. Department of Defence, through transformation of the U.S. Armed Forces and global force posture realignments, is maintaining the capacity to defend against Beijing's use of force or coercion against Taiwan.

Beijing's Strategy in the Taiwan Strait

Beijing appears prepared to defer the use of force as long as it believes the trend of cross-Strait relations continues toward unification and the costs of a conflict outweigh the benefits. In the near term, Beijing's objective appears to be focused on preventing Taiwan from moving toward de jure independence through a strategy that integrates political, economic, cultural, legal, diplomatic, and coercive military instruments of power. Although Beijing professes a desire for peaceful unification that would allow Taiwan to retain a high degree of autonomy, the PLA's deployment of SRBMs, enhanced amphibious warfare capabilities, and modern, advanced long-range anti-air systems across the Strait from Taiwan underscores that Beijing remains unwilling to renounce the use of force.

The circumstances under which the Mainland has historically warned it would use force have evolved over time in response to the island's declarations of political status, changes in PLA capabilities, and Beijing's view of Taiwan's relations with other countries. These circumstances, or "red lines," have included:

a) Formal declaration of Taiwan independence;

b) Undefined moves toward Taiwan independence;

c) Internal unrest on Taiwan;

d) Taiwan's acquisition of nuclear weapons;

e) Indefinite delays in the resumption of cross-Strait dialogue on unification;

f) Foreign intervention in Taiwan's internal affairs; and,

g) Foreign troops stationed on Taiwan.

Furthermore, Article 8 of the March 2005 "Anti-Secession Law" states that Beijing may use "non-peaceful means" if "secessionist forces... cause the fact of Taiwan's secession from China;" if "major incidents entailing Taiwan's secession" occur; or if "possibilities for peaceful reunification" are exhausted. The ambiguity of these "red-lines" enhances the credibility of Beijing's deterrence and allows it the flexibility to determine the nature, timing, and form of its response. Added to this ambiguity are political factors internal to the regime that could affect Beijing's decision calculus.

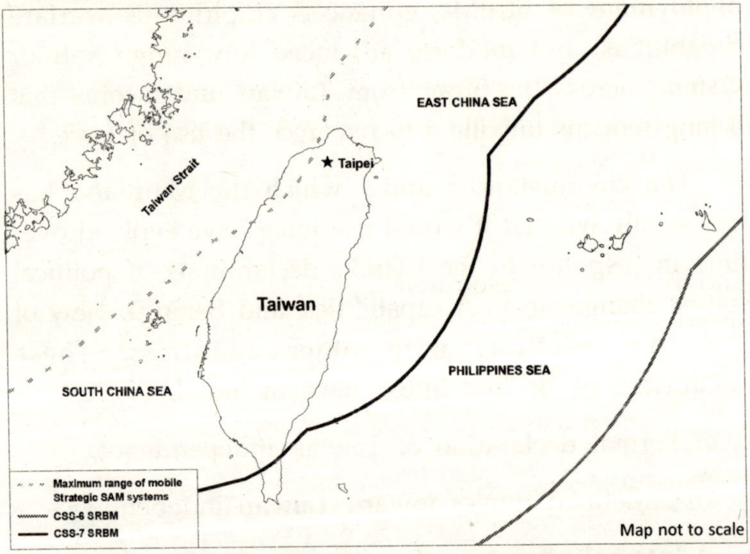

Taiwan Strait SAM & SRBM Coverage. *This map depicts notional coverage based on the range of the Russian-designed SA-20 PMU2 SAM system and the CSS-6 and CSS-7 SRBMs. Actual coverage would be non-contiguous and dependent upon precise deployment sites. If deployed near the Taiwan Strait, the PMU2's extended range provides the PLA's SAM force with an offensive capability against Taiwan aircraft.*

PLA Planning for Taiwan Contingencies

The concepts detailed in Chapters Two and Three would shape the PLA's use of military force against Taiwan. In any contingency, Beijing would face the dual challenge of degrading Taiwan's will to resist while deterring or countering intervention by third parties. Numerous PRC statements describe the United States as the most likely third country to intervene, as well as the most difficult military force to counter. Therefore, it is logical that Beijing would require its military planners to assume and address U.S. military intervention in any Taiwan Strait contingency.

Beijing's Courses of Action Against Taiwan

The PLA is capable of increasingly sophisticated military action against Taiwan. Some analysts hold that Beijing would first pursue a measured approach characterized by signaling its readiness to use force, followed by deliberate buildup of force to optimize the speed of engagement over strategic deception. Others contend that it is more likely that Beijing would sacrifice preparations in favour of surprise to force rapid military and/or political resolution before other countries could respond. If a quick resolution is not possible, Beijing would seek to:

a) Deter potential U.S. intervention;

b) Failing that, delay intervention and seek victory in an asymmetric, limited, quick war; or,

c) Fight to a standstill and pursue a political settlement after a protracted conflict.

Maritime Quarantine or Blockade: Although a traditional maritime quarantine or blockade would have greater impact on Taiwan, it would also tax PLA Navy capabilities. PRC military writings describe potential alternative solutions - air blockades, missile attacks, and mining - to obstruct harbours and approaches. Beijing could declare that ships en route to Taiwan must stop in mainland ports for inspection prior to transiting to Taiwan ports. Beijing could also attempt the equivalent of a blockade by declaring exercise or missile closure areas in approaches to ports, in effect closing port access and diverting merchant traffic. The PLA employed this method during the 1995- 96 missile firings and live-fire exercises. There is risk, however, that Beijing would underestimate the degree to which any attempt to limit maritime traffic to and from Taiwan would trigger countervailing international pressure and military escalation.

Limited Force or Coercive Options: Beijing might use a variety of disruptive, punitive, or lethal military actions in a limited campaign against Taiwan, likely in conjunction with overt and clandestine economic and political activities. Such a campaign could include computer network or limited kinetic attacks against Taiwan's political, military, and economic infrastructure to induce fear on Taiwan and degrade the populace's confidence in the Taiwan leadership. Similarly, PLA special operation forces that have infiltrated Taiwan could conduct attacks against infrastructure or leadership targets.

Air and Missile Campaign: Limited SRBM attacks and precision strikes against air defence systems, including air bases, radar sites, missiles, space assets, and communications facilities, could support a campaign to degrade Taiwan's defences, neutralize Taiwan's military and political leadership, and possibly break the Taiwan people's will to fight.

Amphibious Invasion: Publicly available PRC writings describe different operational concepts for amphibious invasion. The most prominent of these, the Joint Island Landing Campaign, envisions a complex operation relying on coordinated, interlocking campaigns for logistics, air and naval support, and electronic warfare. The objective would be to break through or circumvent shore defences,. establish and build a beachhead, transport personnel and material to designated landing sites in the north or south of Taiwan's western coastline, and launch attacks to seize and occupy key targets and/or the entire island.

The PLA is capable of accomplishing various amphibious operations short of a full-scale invasion of Taiwan. With few overt military preparations beyond routine training, China could launch an invasion of small Taiwan-held islands such as the Pratas or Itu Aba. This invasion would demonstrate military capability and political resolve while achieving tangible territorial gain and simultaneously showing some measure of restraint. However, this kind of operation includes significant, if not prohibitive, political risk because it could galvanize the Taiwan populace and generate international opposition. A PLA invasion of a medium-sized, defended offshore island such as Mazu or Jinmen is within China's capabilities.

Large-scale amphibious invasion is one of the most complicated and difficult military manoeuvres. Success depends upon air and sea superiority, rapid buildup and sustainment of supplies on shore, and uninterrupted support. An attempt to invade Taiwan would strain China's untested armed forces and invite international intervention. These stresses, combined with China's combat force attrition and the complexity of urban warfare and counterinsurgency (assuming a successful landing and breakout), make amphibious invasion of Taiwan a significant political and military risk. Taiwan's investments to harden infrastructure and strengthen defensive capabilities could also decrease Beijing's ability to achieve its objectives.

Factors of Deterrence

Several factors may deter China from taking military action against Taiwan. First, China does not yet possess the military capability to invade and conquer the island, particularly when confronted with the prospect of U.S. intervention. Moreover, an insurgency directed against any PRC presence could tie up PLA forces for years. A military conflict in the Taiwan Strait would also affect the interests of Japan and other nations in the region that advocate a peaceful resolution of the cross-Strait dispute, and would likely result in a fundamental re-ordering of the East Asian security architecture.

Beijing would also have to factor in its calculus the potential political and economic repercussions of a military conflict with Taiwan. China's leaders recognize that a war could severely retard economic development. Taiwan is China's single largest source of foreign direct investment, and an extended campaign would wreck Taiwan's economic infrastructure, leading to high reconstruction costs. International sanctions could further damage Beijing's economic development. A conflict would also severely damage the image that Beijing has sought to project and would undermine the goodwill Beijing has attempted to build through its "Harmonious World" campaign leading up to and following its successful hosting of the 2008 Summer Olympic Games. A conflict could also trigger domestic unrest on the mainland, a contingency that Beijing appears to have factored into its planning. Finally, China's leaders recognize that a conflict over Taiwan involving the United States would lead to a long-term hostile relationship between the United States and China - a result that would not be in China's interests.

CHAPTER 6

Annual Update

"The future and destiny of contemporary China is more and more closely linked to the future and destiny of the world. China's development cannot be done without the world, and the world's development needs China."

- PRC President Hu Jintao

Several significant developments in China over the past year relate to the questions Congress posed in Section 1202 of the National Defence Authorization Act for Fiscal Year 2000 (Public Law 106-65).

Developments in China's Strategy

- China's State Council published the White Paper on China's National Defence in 2008 on January 20, 2009. Consistent with previous editions of the White Paper, the 2009 version exhibits limited improvements in transparency and openness.
- China refused to support Russia's effort to use the August 2008 SCO Summit in Dushanbe to endorse Moscow's invasion of Georgia and subsequent recognition of the Georgian provinces of South Ossetia and Abkhazia.

- From August 8-22, 2008, China hosted the Olympic Games, enhancing China's international prestige and strengthening domestic support for the CCP. To maintain security throughout the Games, the PLA dispatched 87,000 police officers and 12,000 police vehicles in Beijing and Tianjin from July 21 to August 25. The PLA also deployed 34,000 soldiers, 122 military aircraft, and 33 ships, as well as surface-to-air missiles and radars. Militia and PLA reserve units mobilized to assist in providing security for event venues.

- In December 2008, the PLA Navy deployed two destroyers and one supply ship to the Gulf of Aden to conduct counter-piracy escort and patrol operations. Outside of occasional ship visits, this represents the PLA Navy's first operational deployment beyond the immediate Western Pacific region.

- In March 2008, the United States and China completed installation of a Defence Telephone Link (DTL), establishing direct communications between the Department of Defence and China's Ministry of National Defence (MND). Secretary Gates completed the first DTL call to PRC Minister of National Defence Liang Guanglie in April 2008.

Developments in the Security Situation in the Taiwan Strait

- Since the election of President Ma Ying-jeou in March 2008, the security situation in the Taiwan Strait has entered a period of relaxing tensions. Both

Beijing and Taipei have emphasized enhancing people-to-people contacts and expanding economic ties. However, to date, there have been no meaningful actions on the part of the Mainland to reduce its military presence directly opposite Taiwan.

- Preceding the March 2008 Taiwan Presidential election, PRC rhetoric appeared to signal a greater willingness by Beijing to consider the use of limited force to prevent independence. Although tensions decreased after the election of Ma Ying-jeou, there have been no signs that Beijing's military dispositions opposite Taiwan have changed significantly.

- In a statement after the March election on Taiwan, PRC President Hu Jintao proposed that the Mainland and Taiwan "build mutual trust, lay aside disputes, seek consensus and shelve differences, and create a win-win situation" to secure peace and promote the "peaceful development of cross-Strait relations." In a signal of his openness to the overture, newly elected Taiwan President Ma referenced Hu's statement during his inaugural address.

- In June 2008, Taiwan's Straits Exchange Foundation (SEF) and China's Association for Relations Across the Taiwan Strait (ARATS) reinstituted semi-official bilateral exchanges after a nine-year hiatus. The first round of dialogue centered on noncontroversial issues: regular direct weekend cross-Strait charter flights and increased mainland tourism to Taiwan. The dialogue did not address sensitive political subjects such as sovereignty and Taiwan's international space.

- ARATS Chairman Chen Yunlin visited Taiwan in November 2008 to sign agreements on cross-Strait flights, direct maritime shipping, mail service, and food safety with his SEF counterpart, P.K. Chiang. The main opposition party on Taiwan, the Democratic Progressive Party, protested against Chen's visit to Taiwan.
- On December 31, 2008, PRC President Hu Jintao gave a speech commemorating the 1979 "Message to Compatriots in Taiwan" in which the PRC abandoned the idea of "armed liberation" in favour of "peaceful liberation." Hu's speech outlined six points that would govern future cross-Strait relations including cross-Strait military confidence-building measures and negotiation of a peace agreement.

Developments in the Size, Location and Capabilities of PRC Military Forces

China's long-term, comprehensive transformation of its military forces is improving its capacity for force projection and anti-access/area-denial. Consistent with a near-term focus on preparing for Taiwan Strait contingencies, China continues to deploy many of its most advanced systems to the MRs opposite Taiwan. China describes operating under "informatized" conditions, improving "integrated joint operations," and preparing for warfare in a "complex electromagnetic environment" as the primary objectives for the PLA's build-up.

- On May 12, 2008, an earthquake measuring 8.0 on the Richter scale struck Sichuan province in southwestern China. The death toll likely exceeded

80,000 and severely damaged infrastructure, including roads and dams. China quickly mobilized rescue and relief efforts that included deployment of many senior civilian and military leaders and disaster management personnel to affected areas.

- According to PRC press, 20,000 military and armed police had arrived in Sichuan within ten hours of the earthquake, and eventually 139,000 PLA and PAP troops were involved in rescue and recovery efforts. Although the PLA demonstrated the ability to deploy thousands of troops to the affected regions quickly, it had difficulty providing logistics support, leaving many troops in the field without the necessary equipment and supplies to conduct an effective disaster relief operation. Nearly 50,000 militia and PLA reserve forces also supported relief efforts. Militia and reserve units transported supplies into and evacuated the wounded from the disaster area.

- In January and February 2008, Beijing mobilized military assets, including militia and reserve units, to help with relief during a snow storm that crippled large swaths of eastern China. Newly appointed Nanjing Military Region Commander Lieutenant General Zhao Keshi characterized the effort as similar in nature to a wartime environment with a multi-tiered command system and sustained joint logistics.

Ballistic and Cruise Missiles: China has the most active land-based ballistic and cruise missile programme in the world. It is developing and testing offensive missiles, forming additional missile units, qualitatively upgrading certain missile systems, and developing methods to counter ballistic missile defences.

- The PLA is acquiring large numbers of highly accurate cruise missiles, such as the domestically produced ground-launched DH-10 LACM; the domestically produced ground- and ship-launched YJ-62 ASCM, the latter of which is outfitted on the domestically produced LUYANG II-class DDGs; the Russian SS-N-22/SUNBURN supersonic ASCM, which is outfitted on China's SOVREMENNYY-class DDGs and acquired from Russia; and the Russian SS-N-27B/SIZZLER supersonic ASCM, which is outfitted on China's Russian-built KILO-class diesel electric submarines.

- The PLA recently completed an upgrade to the ground-launched YJ-62 ASCM. The new variant, the YJ-62C, has a range of more than 150 NM. According to press reports, the PLA Navy has deployed 120 YJ-62Cs to naval bases in Fujian province, across from Taiwan.

- By September 2008, the PLA had deployed between 1,050 and 1,150 CSS-6 and CSS-7 SRBMs to units opposite Taiwan. It is increasing the size of this force at a rate of more than 100 missiles per year, including variants of these missiles with improved ranges, accuracies and payloads.

- China is developing an ASBM based on a variant of the CSS-5 MRBM as a part of its anti-access strategy.

The missile has a range in excess of 1,500 km, is armed with a manoeuvreable warhead, and when incorporated into a sophisticated command and control system, is intended to provide the PLA the capability to attack ships at sea, including aircraft carriers in the western Pacific Ocean.

- China is modernizing its longer-range ballistic missile force by adding more survivable systems. Most notably, the road-mobile, solid-fueled, nuclear-capable DF-31 ICBM was deployed in 2006, and the longer-range DF-31A was deployed in 2007. The DF-31A, with a range of 11,200 km, can target any location in the continental United States (CONUS).
- China is also developing a new SLBM, the JL-2, for eventual deployment aboard up to five new JIN-class (Type 094) SSBN, which the Department of Defence assesses will achieve Initial Operational Capability (IOC) by 2009-2010. The missile has a range of at least 7,200 km and would provide China its first credible sea-based nuclear strike capability once it becomes operational.

Naval Forces: China's naval forces include some 75 principal combatants, over 60 submarines, 55 medium and large amphibious ships, and approximately 70 missile-equipped patrol craft.

- China has an active aircraft carrier R&D programme. The PRC shipbuilding industry could start construction of an indigenous platform by the end of this decade. China may be interested in building multiple operational aircraft carriers with support ships in the next decade.

- The PLA Navy has reportedly decided to initiate a programme to train 50 pilots to operate fixed-wing aircraft from an aircraft carrier. The initial programme, presumably land-based, would be followed in about four years by ship-borne training involving the ex-VARYAG, which was purchased by a Chinese company from Ukraine in 1998.

- The PLA Navy is improving its over-the-horizon (OTH) targeting capability with Sky Wave and Surface Wave OTH radars, and is developing missiles with improved range and accuracy. OTH radars improve long-range targeting and could be used in conjunction with overhead imagery from satellites to assist in locating targets at great distances from PRC shores for its next generation anti-ship missiles.

- Two new SHANG-class (Type 093) nuclear powered attack submarines (SSN) and one JIN-class (Type 094) SSBN have entered service alongside four older HAN-class SSNs and China's single XIA-class SSBN.

- China has an estimated 13 SONG-class (Type o 039) diesel-electric attack submarines (SS) in its inventory. The SONG-class SS is designed to carry the YJ-82 (CH-SS-N-7) ASCM. The follow-on to the SONG is the YUAN-class SS. The first unit is already in fleet service and the second is conducting sea trials. Based on German diesel engine purchases, China may plan to construct 15 additional YUAN-class submarines. The YUAN-class SS are armed similarly to the SONG-class SS, but also include a possible air independent propulsion (AIP) system.

- Press reported on construction of a new PLA Navy base on Hainan Island. The base appears large enough to accommodate a mix of attack and ballistic missile submarines and advanced surface combatant ships. The port, which has underground facilities, would provide the PLA Navy with direct access to vital international sea lanes, and offers the potential for stealthy deployment of submarines into the deep waters of the South China Sea.

- The PLA Navy recently received several new domestically produced surface combatants. These include two LUYANG II-class (Type 052C) DDGs fitted with the indigenous HHQ-9 long-range surface-to-air missile (SAM); two LUZHOU-class (Type 051C) DDGs equipped with the Russian SA-N-20 long-range SAM; and four JIANGKAI II-class (Type 054A) guided missile frigates (FFG) to be fitted with the medium-range HHQ-16 vertically launched naval SAM currently under development. These ships reflect the leadership's priority on an advanced anti-air warfare capability for China's naval forces, which has historically been a weakness of the fleet.

- China is continuing construction of its new HOUBEI-class (Type 022) wave-piercing catamaran hull missile patrol boat. More than 40 of these units have already entered service. Each boat can carry up to eight YJ-83 ASCMs.

- China launched the 10,000-ton ANWEI-class o hospital ship in October of 2008. In addition to providing the the PLA with an at-sea medical capability, the ANWEI-class may help support Humanitarian Assistance/Disaster Relief (HA/DR) efforts in Asia.

- In May 2008, PRC naval militia forces (commercial fishing vessels, referred to as a Militia Offshore Support Detachment) purportedly refueled and supplied two PLA Navy warships operating at sea in a designated sea area off the coast of Zhejiang. The militia provided ammunition and other stores in addition to fuel. The extent to which Militia Offshore Support Detachments could provide logistics and sustainment support to PLA Navy vessels at distances from China remains unknown.

Air and Air Defence Forces: China bases 490 combat aircraft within un-refueled operational range of Taiwan, and has the airfield capacity to expand that number by hundreds. Many of these aircraft are upgrades of older models; however, newer and more advanced aircraft make up a growing percentage of the inventory. Although the PLA Navy could use aircraft to strike surface ships, it lacks the numbers of aircraft and operational ranges necessary for aerial combat over blue water. The vast majority of PLA Navy tactical aircraft for aerial combat consists of older F-7/FISHBED and F-8II/FINBACK interceptors. These interceptors do not possess the range to conduct aerial combat far from their airfields. To conduct long-range over water combat, the PLA Navy has one Su-30MK2/FLANKER regiment consisting of just 23 aircraft.

- The modernized FB-7A fighter-bomber augments other multi-role and strike aircraft, such as the F-10 and Su-30MKK, already deployed with China's air forces.

- China is upgrading its B-6 bomber fleet (originally adapted from the Russian Tu-16) with a new variant which, when operational, will be armed with a new long-range cruise missile.
- The PLAAF has received eight battalions of upgraded Russian SA-20 PMU-2 long-range (200km) SAM systems since 2006. The SA-20 system reportedly provides limited ballistic and cruise missile defence capabilities. Russian press reporting suggests another eight battalions could be delivered within the next two years.
- China's aviation industry is developing several types of AEW&C aircraft. This includes the KJ-200, based on the Y-8 transport, for AEW&C as well as intelligence collection and maritime surveillance, and the KJ-2000, based on the Russian A-50 airframe.
- Press reporting indicated that China remains interested in procuring Su-33 carrier-borne fighters from Russia. Additionally, Russia and China reportedly re-started negotiations for China's purchase of 34 IL-76 transport aircraft and four IL-78 aerial refueling tankers. While the two sides agreed to the sale for $1.045 billion in 2005, Russia has refused to deliver the planes, citing higher manufacturing costs and concern that China's defence industries may reverse-engineer the aircraft for indigenous production. Russia is attempting to increase the contract to $1.5 billion. China has no

other source for large aircraft to augment Chinese military AEW&C capabilities.

Ground Forces: The PLA has about 1.25 million personnel in its ground forces, with approximately 400,000 based in the three MRs opposite Taiwan. China is upgrading these units with modern tanks, armored personnel carriers, and artillery. Among the new capabilities acquired by PLA ground forces are the approximately 200 Type 98 and Type 99 third generation main battle tanks primarily deployed to units in the Beijing and Shenyang MRs, a new generation amphibious assault vehicle (AAV), a 130 km-range 12-tube 200-mm multiple rocket launch system, and the 6-tube 400-mm multiple rocket launcher system reportedly capable of a 200-km-range.

Developments in PLA Doctrine

- In July 2008, the PLA promulgated the 7th revision of the Outline of Military Training and Evaluation (OMTE), the authoritative guide to how the PLA organizes, implements, and evaluates training. In 2008, over 150 PLA units were involved in testing and validating the draft OMTE, which became standard across the entire force in early 2009. The new OMTE emphasizes realistic training conditions, training in electromagnetic and joint environments, and integrating new and high technologies into the force structure. The latest revision codifies new training requirements for anti-terrorism operations, HA/DR, and international peacekeeping.
- As part of the PLA's emphasis on improving the combat readiness through rigorous and realistic

training, the new OMTE also standardizes methods, procedures, and training using opposition forces, referred to as BLUEFOR. BLUEFOR units appear to be semi-permanent, incorporate multiple services and arms, and mimic foreign tactics and command procedures.

- Drawing on analysis of U.S. operations in Iraq and Afghanistan, and internal PRC studies of informatization, the PLA is debating changes in ground force combat models. The PLA traditionally defined the basic pattern of ground war as massive attacks aimed at destruction and attrition. A new model, using "information-plus-firepower," considers ground forces as integrated within a joint force focused on rapid occupation of key strategic targets and stabilization of the battlefield. A PLA publication stated that, "rapid occupation and stable control have become the basic role of the army in operations in the information age."

- September and October 2008 respectively, the PLA General Staff Department (GSD) organized two exercises, LIBING-2008 and LIANHE-2008, each of which involved elements from different MRs. Training between MRs is unusual and highlights PLA efforts to improve mobility and enhance training realism by forcing units to operate on unfamiliar terrain. Both exercises also emphasized command training necessary for effective combined-arms and joint operations, as stipulated in the new OMTE.

Developments in PRC efforts to develop, acquire, or gain access to Advanced Technologies that could enhance its Military Capabilities

- According to a 2008 FBI statement, PRC intelligence services "pose a significant threat both to the national security and to the compromise of U.S. critical national assets." This statement concludes that these services "will remain a significant threat for a long time." The U.S. intelligence community noted that, of all foreign intelligence organizations attempting to penetrate U.S. agencies, China's are the most aggressive.
- Shu Quansheng, a naturalized U.S. citizen who worked as a physicist in the United States, pleaded guilty to violating the Arms Export Control Act by providing the PRC with information on the design and development of a fueling system for space launch vehicles.
- Chi Mak, a PRC national, acknowledged being placed in the United States for more than 20 years to conduct espionage against the United States, providing sensitive plans for U.S. Navy ships, submarines, and weapons to the PRC. In March 2008, he was sentenced to twenty-four and a half years in prison by a federal judge.

Challenges to Taiwan's Deterrent Forces

There were no armed incidents in the vicinity of the Taiwan Strait in 2008 and the overall situation remains stable, as it did in 2007. However, China's military build-up and the deployment of advanced capabilities opposite the island have not eased.

- Taiwan recently reversed the trend of the past several years of declining defence expenditures; it is also modernizing select capabilities and improving its overall contingency training. The balance of forces continues, however, to shift in the Mainland's favour.
- Taiwan continues to bolster its defence by strengthening its crisis management structure, instituting military personnel reforms, improving its joint capabilities and modernizing its equipment.
- Consistent with the provisions of the Taiwan Relations Act, Public Law 96-8 (1979), the United States continues to make available defence articles, services, and training assistance to enable Taiwan to maintain a sufficient self-defence capability.

Developments in China's Asymmetric Capabilities

Space and Counterspace Capabilities: China is rapidly improving its space-based intelligence, surveillance, reconnaissance, navigation, and communications capabilities, allowing for greater military support from space. In parallel, China is developing a multi-dimensional programme to improve its capabilities to limit or prevent the use of space-based assets by potential adversaries during times of crisis or conflict. Although China's commercial space programme has utility for non-military research, it demonstrates space launch and control capabilities that have direct military application.

- China conducted 11 space launches in 2008, putting 15 satellites in orbit. Included in this number are four new remote sensing satellites: Yaogan-4, Yaogan-5, Huanjing-1A, and Huanjing-1B; the Shenzhou-VII manned spacecraft along with its accompanying small satellite, Banxing-1; three communication satellites; and two meteorological satellites.

- In April 2008, China successfully launched its first data relay satellite, TianLian-1. According to PRC news broadcasts, TianLian-1 was initially tasked to support the launch of Shenzhou-VII manned space mission, increasing surveillance and control coverage of the manned spacecraft's path from 12 percent to roughly 60 percent.

- China began development and testing of the Long March V rocket, the world's largest. Intended to lift heavy payloads into space, it will more than double the sizes of Low Earth Orbit (LEO) and Geosynchronous Orbit (GEO) payloads that China can place into orbit. To support these new rockets, a launch facility near Wenchang on Hainan Island began construction in 2008.

- The Chang'e-1 lunar probe, launched in late 2007, continued to operate successfully with a controlled orbit. Chang'e-2. China plans to land a lunar rover on the moon in 2012.

- China's leaders remain silent about the military applications of China's space programmes and counterspace activities.

Cyberwarfare Capabilities: In 2008, numerous computer systems around the world, including those owned by the U.S. Government, continued to be the target of intrusions that appear to have originated within the PRC. Although these intrusions focused on exfiltrating information, the accesses and skills required for these intrusions are similar to those necessary to conduct computer network attacks. It remains unclear if these intrusions were conducted by, or with the endorsement of, the PLA or other elements of the PRC Government. However, developing capabilities for cyberwarfare is consistent with authoritative PLA military writings on the subject. Publicized 2008 attacks by suspected PRC actors include:

- In April 2008, Indian Government officials confirmed that its Ministry of External Affairs' computer network and servers were the victims of intrusions that appeared to originate in China.
- In May 2008, the Belgian Government reported that it had been targeted by PRC hackers multiple times.
- In May 2008, U.S. authorities investigated whether PRC officials secretly copied contents of a U.S. Government laptop during a visit to China by the U.S. Commerce Secretary and used the information to try to penetrate into commerce computers. The investigation is going on.

Special Topic

China's Global Military Engagement

"Actively participating in international and multilateral affairs and undertaking corresponding international obligations to play a constructive role will be more conducive to the enhancement of China's influencing power in the world."

- Liu Jianfei, Theorist and Scholar, CCP Central Party School

Overview

Since 2002, the frequency and scope of PLA interactions with foreign militaries have increased to include joint exercises, peacekeeping operations, and HA/DR. China's global military engagement seeks to enhance its national power by improving foreign relationships, bolstering its international image, and assuaging concerns among other countries about China's rise. The PLA's global activities also contribute to its transformation through the acquisition of modern weapons, improved operational experience, and professionalization due to access to critical management practices, operational doctrine, and training methods.

The Department of Defence oversees U.S. defence contacts with the PLA to ensure that exchanges and interactions with PRC entities comply with legislative requirements, provide clear benefit to the United States, encourage transparency, consider political and military sensitivities in the region, and mitigate security risks.

Traditional Military Diplomacy

Prior to stepping down in March 2008, General Cao Gangchuan visited at least 24 countries during his five years as China's Minister of National Defence. His replacement, Liang Guanglie, conducted more than 20 high-level meetings with foreign counterparts and delegations in 2008. Such visits and exchanges provide China with opportunities to increase military officers' international exposure, communicate China's positions to foreign audiences, better understand alternative world views, and advance foreign relations through inter-personal contacts and military assistance programmes.

PLA travel abroad, occurring with increased frequency in concert with China's more activist global profile, enables China's military officers to observe and study foreign military command structures, unit formations and operational training. In 2005, the PLA sent 100 delegations abroad. In 2007, a 86-day cruise took a naval task force through three oceans to four European countries. Such diplomatic voyages within and beyond Asia have steadily increased since 2002.

In 2008, the PLA maintained a regular presence in 96 countries with at least 267 attachés posted abroad, compared to 201 attachés posted in 2002 and 220 posted in 2005. In addition to managing daily defence affairs with their respective countries, PRC attachés promote China's defence policies, collect information on political and security affairs, manage security assistance programmes, and arrange and escort visiting delegations.

China also hosts foreign military officers as students in its military academies. In October 2008, nearly 200 foreign military students from over 60 countries observed the PLA exercise QIANFENG 2008, which reportedly involved an armored brigade conducting an offensive manoeuvre in a mountainous area. QIANFENG 2008 was the first PLA exercise that was open to observation by foreign military students.

The number of countries with defence attachés in Beijing is also increasing. In 2008, 94 countries had military attachés in Beijing. This represents a significant increase from 2006, when 79 countries had military attachés in China.

Joint Exercises

PLA participation in bilateral and multilateral military exercises is increasing. The PLA derives political benefit from these exercises both in terms of increased influence and enhanced ties with partner states and organizations. These exercises also contribute to PLA modernization by providing opportunities to improve capabilities in areas such as counterterrorism, mobility operations and logistics. The PLA gains operational insights by observing tactics, command decision making (including commander profiling), and equipment use by more advanced militaries. Such exercises also expose the PLA to military personnel from other countries, who may hold other values or view the world differently.

Many of the PLA's exercises with foreign militaries are categorized as counterterrorism exercises even if they depict scenarios that hold little bearing on actual counterterrorism operations. Beijing has held such exercises bilaterally with Russia, India, Pakistan, Thailand and multilaterally with the SCO and the various countries that participated in AMAN-07.

a) In 2004 and 2006, the PLA participated in exercises in Pakistan, which provided the PLA opportunities to train in a mountainous region outside China.

b) In 2007 and 2008, PLA Special Operations Forces conducted exercises with the Thai military, focusing on hand-to-hand combat and underwater operations.

c) Peace Mission 2005 was a joint exercise held in China with Russia and observers from SCO member and observer states. Participants conducted off-shore blockades, paradrops, airfield seizures and amphibious landings. Russian forces included strategic bombers, and advanced early warning, transport, refueling and fighter aircraft along with modern naval vessels.

d) Peace Mission 2007 was a multilateral exercise held with the SCO in Russia. The exercise established several notable firsts for the PLA, including China's first out-of-country deployment of combat aircraft. Peace Mission 2007 likely tested the PLA's logistics and mobility requirements for long-range deployments.

e) In 2006, China and the United States conducted a two-phased joint maritime search and rescue exercise-one phase in China and the other in the United States. This was the first such exercise between the two militaries.

Between 2002 and 2007, the PLA participated in at least 14 maritime search and rescue exercises with foreign militaries, including two each with Australia, Great Britain, India and Pakistan. These exercises provided opportunities to improve humanitarian support operations, a prominent objective outlined in China's 2006 Defence White Paper.

According to China's 2006 Defence White Paper, the PLA sent observers to military exercises held by Turkey, Thailand, Pakistan, India, Australia, and the United States. PRC analysts concluded from these opportunities that logistics and mobilization are potential vulnerabilities in modern warfare, requiring precisely coordinated transportation and communications. Knowledge gained from observing military planning and execution techniques may lead to improvements in PLA logistics and force planning, as well as improvements to the PLA's ability to cooperate with other countries multilaterally in addressing such issues as humanitarian assistance and disaster relief, UN peacekeeping and other international efforts.

In addition to participating in other nations' exercises, China invited foreign military observers and resident military attachés to observe PLA exercises on at least six occasions since 2003. Inviting foreign military observers to attend PLA exercises enables China to project an overall national image of "peaceful development," and increased transparency.

a) In Warrior-2007, 55 observers from 35 countries, including the United States, witnessed a Shenyang Military Region live-fire exercise.

b) In September 2008, over 110 observers from 36 countries, including the United States, witnessed the culmination of Libing 2008. This was the largest number of foreign visitors ever invited to observe a PLA exercise.

Peacekeeping Operations

Prior to 2002, Beijing limited its participation in UN peacekeeping operations based on its long-stated policy of non-interference in other countries' internal affairs (its participation from 1991-1993 in the UN Transitional Authority in Cambodia being a major exception). Participation in UN peacekeeping operations can serve multiple objectives, including demonstrating support for international stability in troubled regions, improving China's international image, providing opportunities to initiate and expand intelligence collection, and enhancing relationships in the affected areas. Since 2002, China's contributions to UN-sponsored peace operations have climbed to over 1,800 on duty with a total contribution of over 10,000 personnel deployed to 18 missions. The contributions have included engineering, logistics and medical troops, and are evolving to include combat troops and senior officers. China provided several rotations of over 100 police officers to the United Nations Stabilization Mission in Haiti (MINUSTAH).

Beijing also has demonstrated a willingness to deploy personnel on missions where conditions are not completely secure. After the death of a Chinese peacekeeper in Lebanon during the summer of 2006, for example, the PLA increased its troop contributions to the UN Interim Force in Lebanon (UNFIL). In November 2007, Beijing deployed over 300 engineers to support the African Union-UN Mission in Darfur. In June 2007, the PLA conducted its first peacekeeping work meeting to discuss how to effectively organize, train, manage, and rotate China's peacekeepers.

Humanitarian Assistance and Disaster Relief

In the last decade, the PLA increased its participation in international HA/DR missions. Since 2002, the PLA has contributed to ten emergency relief operations in fourteen countries in China's immediate region. PLA involvement in international HA/DR missions appears to be driven by a desire to be perceived as a responsible global power. In 2004, China was criticized for its slow response to the Indian Ocean tsunami and initial pledge of only $2.6 million in aid. Subsequently, China offered a total of over $170 million in assistance, including $19.5 million through the UN. However, it is unclear how much pledged assistance China actually delivered.

According to the Director of the Ministry of National Defence, Foreign Affairs Office, Major General Qian Lihua, PLA peacetime training includes rescue and assistance missions. In May 2008, the PLA deployed its national rescue team to the Sichuan Province in response to a devastating 8.0 magnitude earthquake. PRC press reported that the team experienced significant difficulties due to equipment shortfalls and command and control problems.

The PLA's HA/DR capability remains limited, but China is seeking to collaborate with regional partners to improve its HA/DR capabilities. China and Indonesia drafted the "Association of Southeast Asian Nations (ASEAN) Regional Forum General Guidelines on Disaster Relief Cooperation" to steer the development of Standard Operating Procedures for future HA/DR operations, which were adopted in July 2007. China also proposed a standard operating procedure for disaster relief cooperation for the armed forces of the member states of ASEAN+3.

Arms Sales

Beijing conducts arms sales and training both to enhance foreign relationships and to generate revenue to support its domestic defence industry. China's arms sales range from small arms and ammunition to joint development or transfer of advanced weapons systems. Chinese companies sell primarily to developing countries where China's low-cost weapons sales serve both commercial and strategic purposes. For example, China maintains strong and longstanding military-technical cooperation with Pakistan, which includes both arms sales and defence industrial cooperation. With other countries of strategic importance to China, such as Iran and Sudan, arms sales and other security assistance deepen developing ties and balance PRC energy imports. As PRC arms have become less competitive compared to more sophisticated systems sold by Western or Russian suppliers, arms sales have declined in importance to Beijing as a tool of influence. Nonetheless, arms sales continue to play an important role in China's efforts to influence cash-strapped countries, many of which do not have access to other sources of arms and are willing to trade quality for lower cost. As the quality of PRC arms improves, Beijing may be able to wield arms sales a competitive tool of influence again.

From 2003-2007, China sold nearly $7 billion worth of conventional weapons systems worldwide. China's primary customer for conventional weapons has been Pakistan. Sales to Islamabad include JF-17 aircraft, JF-17 production facilities, F-22P frigates with helicopters, K-8 jet trainers, T-85 tanks, F-7 aircraft, ASCMs, missile technologies, small arms, and ammunition. Sales to other countries include fighter, transport, and jet trainer aircraft, tanks, air defence equipment, rockets, military vehicles, patrol boats, missiles and missile technology and small arms and ammunition.

Protesting disclosure of U.S. weapons sales to Taiwan, China did not provide an official record of its arms sales to the UN Conventional Arms Register (UNCAR) from 1996-2007. However, in September 2007, China's Foreign Ministry announced that Beijing resumed its reporting on conventional arms imports and exports to "demonstrate China's positive attitude on enhancing military mutual trust with various countries in the world."

Arms Sales to Areas of Instability

Several PRC entities continue to provide arms to customers in unstable regions.

a) **Zimbabwe:** In March 2008, South African dockworkers refused to unload a PRC cargo ship carrying 70 tons of small arms and ammunition destined to support President Mugabe's ZANU-PF Party, which is waging a violent campaign of intimidation against pro-democracy advocates. China initially defended the shipment, but with mounting international pressure, off-loaded non-military cargo in Angola, after which the ship returned to China with its remaining military cargo.

b) **Iran:** China supported UN Security Council Resolutions 1737, 1747, 1803, and 1835. However, concerns remain regarding China's enforcement of existing laws regarding arms sales to Iran. A number of transfers to Iran have resulted in U.S. trade penalties and sanctions against entities in China. Some weapons that PRC entities supplied to Iran were found to have been transferred to terrorist organizations in Iraq and Afghanistan. This is a serious issue that the United States continues to monitor.

c) **Sudan:** The PRC government has at times used its influence with the Sudanese government to address in a positive way international concerns over Darfur, yet has also continued to provide political support for Khartoum. China sells arms to Sudan despite the passage of UN Security Council resolutions 1556 (2004) and 1591 (2005), both of which call for the prevention of the transfer of arms to Darfur. The PRC argues that arms sales constitute part of normal commercial relations, and that the arms supplied by Chinese companies were not meant for use in

Darfur. Between 2004 and 2006, China made up an average of 90 percent of small arms sales to Sudan.

As China's regional and international interests expand and grow more complex, the PLA's international engagement will expand, especially in the areas of peace operations, HA/DR, and joint exercises. In addition to furthering PLA modernization, the focus of these engagements will likely remain on building China's political ties, assuaging fears about China's rise, and building China's international influence, particularly in Asia. However, the sale of arms by PRC entities to states of concern and unstable regions are disruptive to regional and global stability.

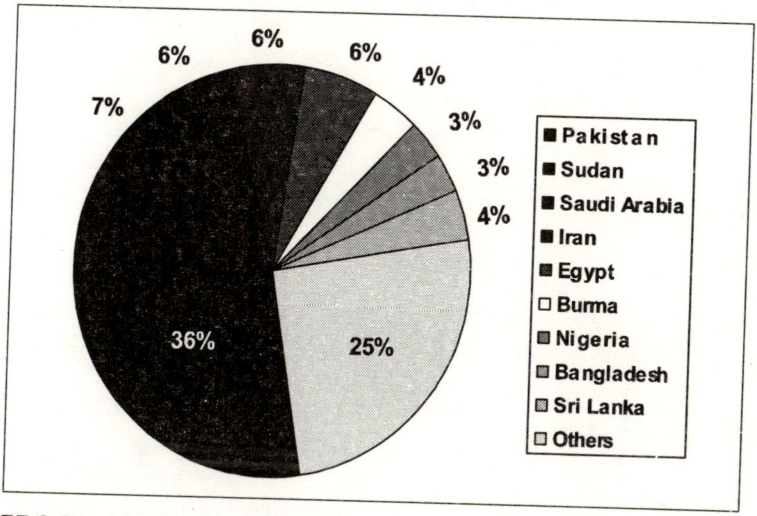

PRC Worldwide Arms Sales Customers, 2003 - 2007. Pakistan was China's largest customer for conventional arms; Sudan was second.

Appendix

China and Taiwan Forces Data

Taiwan Strait Military Balance, Ground Forces

	China		Taiwan
	Total	Taiwan Strait Area	Total
Personnel (Active)	1.25 million	440,000	130,000
Group Armies	18	8	3
Infantry Divisions	19	8	0
Infantry Brigades	24	11	8
Mechanized Infantry Divisions	4	1	0
Mechanized Infantry Brigades	5	1	3
Armor Divisions	9	4	0
Armor Brigades	8	3	5
Artillery Divisions	2	2	0
Artillery Brigades	17	6	5
Airborne Divisions	3	3	0
Amphibious Divisions	2	2	0
Amphibious Brigades	3	3	3
Tanks	6,700	2,800	1,100
Artillery Pieces	7,400	2,900	1,600

Note: PLA active ground forces are organized into Group Armies. Infantry, armor, and artillery units are organized into a combination of divisions and brigades deployed throughout the PLA's seven MRs. A significant portion of these assets are deployed in the Taiwan Strait area, specifically the Nanjing, Guangzhou, and Jinan MRs. Taiwan has seven Defence Commands, three of which have Field Armies. Each Army contains an Artillery Command roughly equivalent to a brigade plus.

Major Ground Units

CHINA: Group Armies (GA) Primary Missions

Shenyang MR
16 GA – Defensive, Offensive CT
39 GA – RRU, Offensive MF
40 GA – Defensive, Offensive CT

Beijing MR
65 GA – Defensive
38 GA – RRU, Offensive MF
27 GA – Defensive

Jinan MR
26 GA – Offensive CT, Defensive
20 GA – Offensive CT, Defensive
54 GA – Offensive MF, Amphibious

Nanjing MR
1 GA – Amphibious, Offensive CT
12 GA – Amphibious, Offensive CT
31 GA – Amphibious, Offensive CT

Guangzhou MR
15 Airborne – RRU, Offensive MF
41 GA – Offensive CT, Amphibious
42 GA – Amphibious

Chengdu MR
13 GA – Defensive, Offensive CT
14 GA – Defensive, Offensive CT

Lanzhou MR
47 GA – Defensive, Offensive CT
21 GA – Offensive MF, Defensive

MR – Military Region
MF – Mobile Force
RRU – Rapid Reaction Unit
CT – Complex Terrain (mountain, urban, jungle, etc)

Taiwan Strait Military Balance, Air Forces

China			Taiwan
Aircraft	Total	Within range of Taiwan	Total
Fighters	1,655	330	390
Bombers/Attack	645	160	0
Transport	450	40	40

Note:The PLAAF and the PLA Navy have approximately 2,300 operational combat aircraft. These consist of air defence and multi-role fighters, ground attack aircraft, fighter-bombers, and bombers. An additional 1,450 older fighters, bombers and trainers are employed for training and R&D. The two air arms also possess approximately 450 transports and over 100 surveillance and reconnaissance aircraftwith intelligence, surface search, and airborne early warning capabilities. The majority of PLAAF and PLA Navy aircraft are based in the eastern half of the country. Currently, 490 aircraft could conduct combat operations against Taiwan without refueling. However, this number could be significantly increased through any combination of aircraft forward deployment, decreased ordnance loads, or altered mission profiles.

Major Air Units

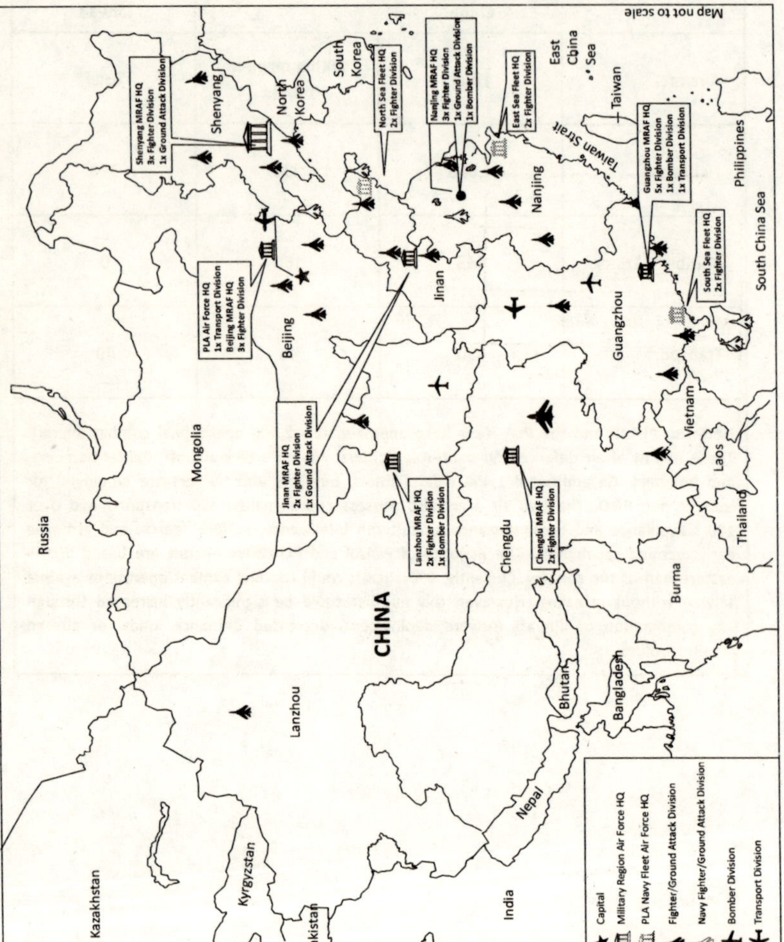

Taiwan Strait Military Balance, Naval Forces

	China		Taiwan
	Total	East and South Sea Fleets	Total
Destroyers	27	17	4
Frigates	48	39	22
Tank Landing Ships/ Amphibious Transport Dock	27	25	12
Medium Landing Ships	28	23	4
Diesel Attack Submarines	54	32	4
Nuclear Attack Submarines	6	1	0
Coastal Patrol (Missile)	70	55	59

Note: The PLA Navy has the largest force of principal combatants, submarines, and amphibious warfare ships in Asia. After years of neglect, the force of missile-armed patrol craft is also growing. In the event of a major Taiwan conflict, the East and South Sea Fleets would be expected to participate in direct action against the Taiwan Navy. The North Sea Fleet would be responsible primarily for protecting Beijing and the northern coast, but could provide mission-critical assets to support other fleets.

Major Naval Units

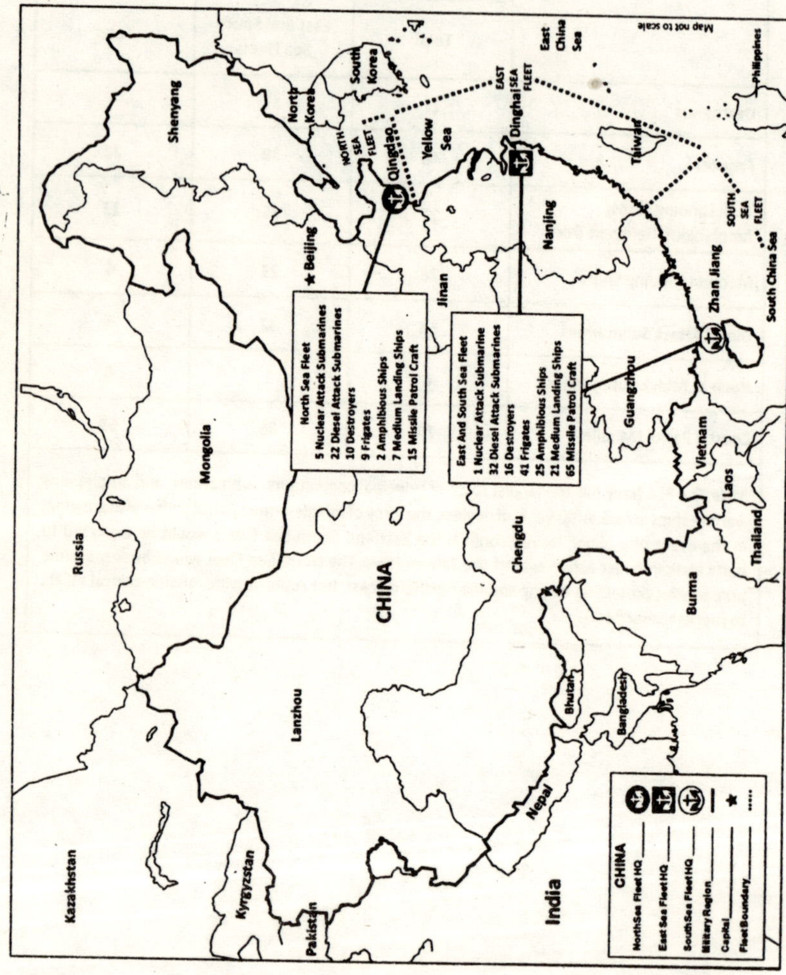

Inventory of PLAAF Surface-to-Air Missile Launchers

System	2008
SA-20 (S-300PMU2)	32
SA-20 (S-300PMU1)	64
SA-10B (S-300PMU)	32
HQ-9	64
KS-1A	60
HQ-6	30
CSA-1 and variants	400

China's Missile Forces

China's Missile Inventory	Ballistic and Cruise		Estimated Range
	Missiles	Launchers	
CSS -2	15-20	5-10	3,000+ km
CSS -3	15-20	10-15	5,400+ km
CSS -4	20	20	13,000+ km
DF -31	<10	<10	7,200+ km
DF -31A	<10	<10	11,200+ km
CSS -5	60-80	70-90	1,750+ km
CSS -6	350-400	90-110	600 km
CSS -7	700-750	120-140	300 km
DH -10	150-350	40-55	1,500+ km
JL -2	Developmental	10-14	7,200+ km

Note: China's Second Artillery maintains at least 5 operational SRBM brigades; an additional two brigades are subordinate to PLA ground forces –one garrisoned in the Nanjing MR and the other in the Guangzhou MR. All SRBM units are deployed to locations near Taiwan.

Index

Afghanistan, 42, 117
Africa, 29
Air independent propulsion (AIP) system, 112
Airborne Early Warning and Control (AEW&C), 74
Aircraft, F-10 fighter, 81
Air-to-Surface Missiles (ASMs), 64
AMAN-07, 126
Amphibious Assault Vehicle (AAV), 116
Angola, 28, 29, 132
Anti-Radiation Missiles (ARM), 65
Anti-Secession Law, March 2005, 98
Anti-ship Cruise Missiles (ASCMs), 17, 64
Armed Anti-ship Ballistic Missiles (ASBMs), 61
Arms, quality of PRC, 131
Arunachal Pradesh, 30
Ascent Anti-Satellite (ASAT), 48
ASCM, SS-N-27B/SIZZLER Supersonic, 110
Asia
  democratic revolutions in Central–36
  energy investment in Central–74
  military operations, 73
"Assassin's Mace", 60
Association of Southeast Asian Nations (ASEAN), 130
Australia, 28, 127

B-6 bomber fleet, 115
Banxing-1(Bx-1), 69, 120
Beijing, 24, 36, 59, 74, 106, 126, 130
Bei Dou, System-1, 2, 69
BLUEFOR, 117
Budget, China's military, 78

C4ISR, 19, 55, 62, 72, 92
  China's space-based, 46
    improvement in China's, 74
CCP, 38, 106
Central Asia, 76
Central Military Commission (CMC), 19, 41
Chad, 28
Chang'e-2, 120
Chang'e-1, first lunar orbiter, 69, 120
Chen Yunlin, ARATS Chairman, 108
Chi Mak, 118
China Military Science, 46
China, 15, 16, 17, 18, 19, 20
  actual policy decisions, 21
  corruption in, 34
  Mainland, 36
  military attaches in, 125
  reforms have enabled, 25
China's 2006 Defence White Paper, 127
China's Association for Relations Across the Taiwan Strait (ARATS), 107
China's doctrine, 52
China's Ministry of National Defence (MNL), 106
China's National Defence University, 42
China's SOVREMENNYY class DDGs, 110
Chinese Communist Party (CCP), 24
Chunxiao/Shirakaba gas field, 29
Civil communications satellites (COMSATs), 70
CMC, 49
Committee, CCP Central, 49
Comprehensive national power (CNP), 23
Computer network attack (CNA), 62, 72,
Computer Network Exploitation (CNE), 72
Computer Network Operations (CNO), 72
Community, U.S. Intelligence, 60
Congress, 2012 18th Party, 38
Congress, CCP's 17th Party, 83
CSS-2, 66
CSS-3 ICBMs, 66
CSS-4 ICBMs, 66
CSS-5, 66
CSS-5 MRBM, 110

Darfur, 132

Darfur, African Union-UN Mission in, 129
Datang, 85
DDGs, LUYANG II-class (Type 052C), 113
DDGs, SOVERMENNYY II class, 62
Defence Export Control Act, 88
Defence Telephone Link (DTL), 106
Democratic Progressive Party, 108
Demonstrators, crackdown on Tiananmen Square, 88
Deng Xiaoping, 22, 37
DF-31, 16
DF-31 ICBM, 111
DF-31A, 16
DH-10, 64
DH-10 LACM, 110
Diesel-electric attack submarines (SS) 112
Dushanbe, August 2008 SCO Summit in, 105

Earthquake, Sichuan, 51, 78
East China Sea, 29, 74
Egypt, 28
Electronic Warfare (EW), 18
European Union (EU), 88
Exclusive Economic Zones (EEZs), 29

Federal Bureau of Investigation (FBI), 89
FFG, JIANGKAI II-class (Type 054A) guided missile frigates, 113
Forces, PLA Reserve, 45
Foreign policy, "white paper" on China's–21
  arms control, 21
  national defence, 21
Fujian province, 110

General Cao Gangchuan, 124
Georgia, Moscow's invasion of, 105
Geosynchronous Orbit (GEO), 120
Global positioning system (GPS), US 69
GLONASS, Russia's, 69
Gobi desert, 36
Great Britain, 127
Gulf of Aden, 106

Hainan Island, 120
  PLA Navy base on, 113
HARPY, Israeli made, 65

Hu Jintao, 55
  PRC President, 77, 83, 105, 107, 108
Hu's "Harmonious World" 22
Huang Xueping, 93
Huanjing-1B, 120
Huawei, 85

India, 126, 127
  border conflicts against, 44
  China's border conflicts with, 29
  China's shared border with, 75
Indonesia, 28
Industry, Armament, 87
Industry, Aviation, 87
Industry, Missile and Space 86
Industry, Shipbuilding, 86
Information Technology Companies, 85
  Datang, 85
  Huawei, 85
  Zhongxing, 85
Interceptors, F-7/FISHBED and F-8II/FINBACK, 114
Intercontinental-range Ballistic Missiles (ICBMs), 16
Iran, 29, 130
  laws regarding arms sales to, 132
Iraq, 17, 42
Islamabad, sales to, 131
  Island, Paracel, 30
  Island, Spratly, 30
Islands, invasion of small Taiwan-held, 101
Islands, Paracel, 74
Israel, 88
Israeli HARPY, 63
Itu Aba, 30, 101

Japan, 28, 30, 36, 37
  Disagreements with, 74
JF-17 aircraft, 131
Jiang Zemin, 41
JIN-class (Type 094) SSBN, 111, 112
Jinmen, 101
Joint Space War Campaigns (2005), 47
JOINT 2007, 51

Kazakhstan, 28, 31, 76
Khartoum, 132
Korean Peninsula, disruptions on the, 36

Land Attack Cruise Missile (LACM), 18, 62, 64

Lebanon, Chinese peacekeeper in, 129
LIANHE-2008, 117
Liberation Army Daily, 71
LIBING-2008, 117, 128
Liang Guanglie, 124
  PRC Minister of National Defence, 106
Liu Jianfei, 123
Long March V rocket, 120
Low Earth Orbit (LEO), 120
LUYANG II-class DDGs, 110

Ma Ying-jeou, Taiwan President, 36, 96, 106
Mainland China, 73, 95, 97
Manoeuvring re-entry vehicles (MaRV), 66
Mazu, 101
Medium-Range Ballistic Missiles (MRBMs), 18, 64
Middle East, 28, 29
Military, Secrecy and Deception in PLA, 52
Military, Soviet and Russian, 39
Militia, China's, 45
Milk, China's, 33
Minorities, Tibetan and Uighur, 34
Missiles, AS-17/Kh-31A anti ship, 63
Missiles, ASA, 70
Moon, lunar rover on the, 120
Multiple Independently targeted Re-entry Vehicles (MIRV), 66
Multiple Rocket Launcher (MRL), 65
Munition, Artillery Delivered High Precision, 65

Nanjing Military Region, 109
Nanotechnology, 91
National Defence Authorization Act, 15
National Defence University, 47
National People's Congress, 54
Nigeria, 28, 31, 70
Northeast Asia, 30

Okinawa Trench, 29
Olympics, successful Beijing, 33
Oman, 28
Operation Allied Force, 53

Operation Enduring Freedom, 39
Operation Iraqi Freedom, 39
Organization of Economic Cooperation and Development (OECD), 84
Outline of Military Training and Evaluation (OMTE), 116
Over-the-Horizon (OTH), 112

P.K. Chiang, 108
Pacific Ocean, 74, 111
Pakistan, 126, 127
Paper
 China's, 45
 2004 Defence, White, 45
 2006 Defence, White, 54, 82
 2008 Defence, White, 54, 82
Peace Mission, 2005, 126
Peace Mission, 2007, 126
Peninsula, Instability on the Korean, 74
People's Republic of China (PRC), 21, 30
PLA, 16, 18, 19, 22, 35, 40, 42, 48, 106, 111
PLA Academy of Military Science, 49
PLA Air Force (PLAAF), 18, 46, 56
PLA General Staff Department (GSD), 117
PLA Navy, 42, 113, 19
PLA publication, 50
PLA, U.S. defence contacts with the, 124
PLAAF, 45, 63, 67, 90, 115
Population, China's, 32
Pratas, 101
PRC hackers, 121
PRC, Pranab Mukherjee visited the, 30
Press, China's state-owned, 67

Qian Lihua, Major General, 93, 129
QIANFENG 2008, PLA exercise, 125

Radars, Sky Wave and Surface Wave OTH, 112
Remote sensing satellites,
 Yaogan-4, 120
 Yaogan-5, 120
Russia, 126
Russia, Su-33 carrier-borne fighters from, 115
Russian IL-78, 63
Russian SS-N-22/SUNBURN supersonic ASCM, 110

SAM, Russian SA N-20 long range, 113
Sansha City, 30
Satellite, PRC weather, 70
Satellites, Small, 70
Saudi Arabia, 28, 29
Science of Campaigns (2000), PLA's, 41
Science of Military Campaigns, 50
Science of Military Strategy, 44, 49
Sea power, China, 55
Sea, UN Convention on the Law of the, 49
Secretary Gates, 106
Senkaku Island, 29, 36
Shang Yang, 52
Shanghai Cooperation Organization (SCO), 76
SHARP SWORD 2007, 51
Shenyang Military Region, 128
Shenzhou-VII, 69
Ship, ANWEI-class hospital, 113
Short-Range Ballistic Missiles (SRBMs), 18, 64
Shu Quansheng, 118
Sichuan, 108, 109, 129
Sikkim, 30
SONG-class SS, 112
South China Sea, 26, 30, 74, 74, 113
South Korea, 30, 36, 37
Southeast Asia, 30, 37
Soviet Union, 44
  collapse of the 40
  spacecraft, Shenzhou-VII manned, 120
Spratly Island, 30, 36, 74
SRBMs, CSS-6 and CSS-7, 110
SRBMs, deployment of, 97
SS, YUAN-class, 112
SSBN, China's single XIA-class, 112
  submarine-launched ballistic missile (SLBM) 17
  submarines (SSN), SHANG-class (Type 093) nuclear powered attack, 112
  submarines, China's Russian-built KILO-class diesel electric, 110
  submarines, YUAN-class, 62
Sudan, 28, 29, 31, 130
  China sells arms to, 132

Summit, EU-China, 88
Sun Bin, 52
Sun Zi, 52
Surface-to-air missile (SAM), 63
    such as the HQ-9, SA-10, SA-20, 63

T-85 tanks, 131
Thailand, 126
Tainjin, 106
Taipei, 36, 106
Taiwan, 18, 30, 37, 59, 95
Taiwan Relations Act, Public Law 96-8 (1979), 96, 119
Taiwan Strait, 106, 118
    Beijing's Strategy in the, 97
    missile firings in the, 42
Taiwan, attempt to invade, 102
Taiwan, Beijing's Courses of Action Against, 99
Taiwan, Beijing's objective of unifying, 20
Taiwan, Modernization Beyond, 73
Taiwan's Straits Exchange Foundation (SEF), 109
Tang Jiaxuan, 26
Thai military, 126
Thailand, 126, 127
"Three Warfares," concept of, 49
TianLian-1, data relay satellite, 70, 120
Tibet, suppression operations, 78
Tsunami, Indian Ocean, 129
Turkey, 127
Turkmenistan, 31, 76

U.S. Department of Defence (DoD), 15
U.S. technologies, China's efforts to obtain sensitive; 89
U.S. intervention, 99
Uighur, 76
Ukraine, 112
Ultra High Frequency (UHF), 71
UN Conventional Arms Register (UNCAR), 131
UN Interim Force in Lebanon (UNIFIL), 129
UN peacekeeping, 127
UN Security Council resolutions 1556 (2004), 132
UN Transitional Authority in Cambodia, 128

United Nations (UN), 26, 80
United Nations Stabilization Mission in Haiti (MINUSTAH), 128
United States, 16, 31, 37, 80, 95, 127, 128
Unmanned aerial vehicles (UAVs), 63
Unmanned combat aerial vehicles (UCAVs), 65
Uzbekistan, 76

VARYAG, Soviet Kuznetsov-class aircraft carrier, 93, 112
Venezuela, 29, 70
Vietnam, 44

War, intervention in the Korean, 44
War, Persian Gulf, 40, 53
Warfare, concept of Legal, 49
Warfare, importance of space, 47
Warfare, Integrated Network Electronic, 48
Warfare, Legal, 49
Warfare, Media, 49
Warfare, Psychological, 49
WARRIOR 2007, 51, 128
WARRIOR 2008, 51
Weapons, Anti-Radiation, 65
Weapons, China's low-cost, 130
Wenchang, 120
West Pacific, 73
Western Pacific, 17
White Paper on China's National Defence in 2008, 105
Writings, PRC military, 47
Wu Qi, 52
Xinjiang, Internal security forces in, 76
Xu Caihou, 54

YJ-62 ASCM, 110
YJ-63, 64
YJ-91, 65
Yuan Zelu, Colonel, 47

ZANU-PF Party, President Mugabe's, 132
Zhao Keshi, Lieutenant General, 109
Zhejinag, 114
Zhongxing, 85
Zhou Shengxian, 35
Zimbabwe, 131

Some of our learned Authors

General S Padmanabhan; General VP Malik; Admiral Vishnu Bhagwat; Dr A Sivathanu Pillai, DRDO; Parvez Dewan, IAS; Dhirendra Singh, IAS; Falguni Rajkumar, IAS; G Sundaram, IAS; PS Bhatnagar, IAS; N Vittal, IAS; PK Agrawal, IAS; Sudhir S Baloeria, IAS; NS Sisodia, IAS; Dr Yogendra Narain, IPS; J.K. Bagchi, IAS; TSR Subramanian, IAS; Air Chief Marshal FH Major, Prakash Singh, IPS; DC Nath, IPS; Ajit Doval, IPS; SP Mathur, IPS; Maloy Krishna Dhar, IPS; BR Lall, IPS; PC Dogra, IPS; DC Pathak, IPS; ML Kumawat, IPS; Dr. S Subramanian, IPS; RK Ohri, IPS; RB Sreekumar, IPS; Kalyan K Mitra, IPS; RS Gupta, IPS; K Koshy, IPS; Kuldeep Khoda, IPS; Ajay Kashyap, IPS; Dr N Dilip Kumar, IPS; Dr Nazrul Islam, IPS; Dr DV Guruprasad, IPS; AP Maheshwari, IPS; Loknath Behera, IPS; Dr NC Asthana, IPS; PM Nair, IPS; Dr Sundari Nanda, IPS; Keshav Kumar, IPS; Dr Prateep V Philip, IPS; Muktesh Chander, IPS; VK Agarwal, IRS; Virendra Gupta, IFS; Lt Gen SK Sinha; Lt Gen Raj Kadyan; Lt Gen Mukesh Sabharwal; Lt Gen HB Kala; Lt Gen PC Katoch; Lt Gen Gautam Banerjee; Vice Admiral Suresh Bangara; Vice Admiral Vijay Shankar; Lt Gen Arjun Ray; Lt Gen CK Kapur; Lt Gen NS Narahari; Lt Gen Chandra Shekhar; Lt Gen ML Chibber; Lt Gen Ashok Joshi; Maj Gen Nilendra Kumar; Maj Gen Vinod Saighal; Maj Gen VK Singh; Maj Gen Anjan Mukherjee; Maj Gen Raj Mehta; Maj Gen Samay Ram; Maj Gen Dipankar Banerjee; Maj Gen Y Behl; Air Cmde RV Kumar; Air Marshal SR Deshpande; Air Marshal RC Bajpai; Air Vice Marshal AK Tiwary; Air Vice Marshal Samir K Sen; Air Cmde (Dr) CN Ghosh; Air Cmde NB Singh; Air Cmde RV Phadke; Arvind Gupta, IFS; Ambassador Kishan S Rana, IFS; Cmde Ranjit B Rai; Capt (IN) Alok Bansal; VK Singal, IRS; Sudha Raman; Brig Yash Mohan, VSM; Brig DK Khullar; Brig Trigunesh Mukherjee; Brig. (Dr.) HK Kar; Brig. MS Patial; Brig. Parmodh Sarin; Brig. RP Singh; Brig. Jaspal Singh; Col PK Gautam; Col MN Gulati; Col MS Jaswal; Col Anil Shorey; Lt Col Anil Bhatt; Dr RK Tiwari; Anuj Dhar, Journalist; Prakash Nanda, Journalist; Ranjit Kumar; Vijay Naik, Journalist etc. etc.